The Big Defreeze of the Universe

(Analogies & Thought Experiments)

Copyright © 2019 by Mohammad Hasan Algarhy.

All rights reserved

No part of this book shall be reproduced, stored in a retrieval system, or transmitted by any means, electronic, mechanical, photocopying, recording, or otherwise, without the author's prior written permission. This includes but is not limited to scientific papers, scientific publications, thesis, and books. For any scientific papers or publications based on the new concepts and ideas introduced in this book, the book's author must be contacted and mentioned as one of the authors of the scientific papers or publication.

The Big Defreeze of the Universe

(Analogies and thought experiments)

First Edition 17th of May 2021

Includes bibliography and references

mohammadhasanalgarhy@gmail.com

Chapter 01: The Big Defreeze.

Chapter 02: The straight-line analogy.

Chapter 03: ISWs as strings

Chapter 04: The imaginary complex mother forces.

Chapter 05: The two worlds interpretation.

Chapter 06: The unified complex imaginary forces

Chapter 07: Testability of the ideas introduced in this book

Chapter 08: The main new concepts introduced in this book.

Contents:

Introduction — 11

Chapter 01: The Big Defreeze — 16

The Big Defreeze compatibility — 19

The Big Defreeze and the Bose-Einstein condensates — 24

Laser cooling — 27

Magnetic traps — 27

Interference between two condensates — 28

Vortices and rotating condensates — 29

The rotation of the CMB in a BEC manner — 30

The synthetic material and the optical lattice — 32

Optical trap — 33

Hybrid material — 34

The fermions problem — 34

Chapter 02: The straight-line analogy — 38

The cosmic inflation — 39

The landscape of string theory — 41

Topological manifolds and manifold bundles — 41

The holographic principle — 43

The straight-line analogy — 44

The physics behind the analogy — 49

How does this story fit into the Planck Era? — 50

The Higgs mechanism — 53

Chapter 03: ISWs as strings   55

Strings moving in two dimensions   60

Boundary conditions on the ends of strings   61

Applying Neumann and Dirichlet boundary conditions to the identity stem waves ISWs   63

The tension in strings & ISWs   64

Applying the string tension equations to the ISWs   66

Evolution of information of the identity stem waves ISWs   67

Riemann curvature tensor for the ISWs   73

The Higgs mechanism, ISWs, and spontaneous symmetry breaking   74

Chapter 04: The imaginary complex forces   76

The retrieving force (The R-force)   80

What is a particle?   81

The Spin of massless particles   83

The sleeping muon analogy   83

The potential mathematics of the R-force   85

Introducing the "IN sign" to Schrödinger's equation   85

Energy in shape = A wave function   88

How to relate this to the potential mathematics of the R-force?   90

The Balancing force (The B-force)   91

Resonance   92

The angle of intersection of a wave with its axis   94

Some of the proposed operations of the Balancing force — 99

The balanced state — 99

The balancing force and the cosmological constant — 100

The non-zero principle of the balancing force — 102

The connection coefficients of the balancing force — 102

The balancing force and the fine-tuning — 103

The balancing force as a different perspective — 106

The higher dimension force (The HD force) — 107

The scaling and projection operation of the HD force — 108

Ricci scalar and the scaling and projection operation of the HD force — 112

Conservation and the Higher dimension force (The HD force) — 113

The paradigm shift operations and effects caused by the HD force — 114

Entanglement as a mechanism for the HD force — 115

The mechanism of the HD-force's entanglement — 116

The Notebook thought experiment — 117

What is time in the notebook thought experiment? — 119

The caveman analogy — 121

Chapter 05: The two worlds interpretation — 123

Some of the problems with the many world interpretation — 127

The problem with the quantum measurement — 130

The two worlds interpretation — 132

Where is that second world/version if it is outside the multi-verse space-time? — 134

What laws does it follow if it does not have quantum mechanical behaviors?	135
How does it interact with our world? Does the second version impact the first version? And how?	136
Could the first version impact the second version (its observer version)? And how?	137
Why is an HD-entanglement operation required with our observer version?	138
What would be the source of knowledge/information of your other version in the second world?	139
Why do we need all this loop of encoding information between the two worlds in the first place?	141
What happens to all the other possible scenarios?	141
What would the second version be doing?	144

Chapter 06: The unified complex imaginary forces — 146

The $HDBR_3$ and the four fundamental forces	149
The spectrum of energies of the $HDBR_3$	150
The $HDBR_3$ operations	154
The fundamental differences between the proposed balancing and retrieving forces	156

Chapter 07: Testability of the ideas introduced in this book

Testability of the Big Defreeze	159
The Big Defreeze in the Cosmic Microwave Background	162
Laser cooling, optical lattice, and optical traps for testing the ISWs	166

Testability of the imaginary complex forces	168
Muons for testing the imaginary complex forces	171
The graviton and the imaginary complex forces	174
Testability of the paradigm-shifting operation of the HD-force	175
Chapter 08: The main new concepts introduced in this book	**179**
The idea of the Big Defreeze	180
The identity stem waves (ISWs)	180
Using the Fourier transform to understand the symmetry-breaking mechanism of the ISW crystal	181
The straight-line analogy	181
Dealing with identity stem waves ISWs as strings	182
The imaginary complex forces and their potential properties	182
Introducing the potential operations of the imaginary complex forces and the basics of their potential mathematics	182
The two worlds interpretation	182
The mechanism of encoding information in the two worlds interpretation	183
The mechanism of updating the information of the wave function as per the two world interpretation	183
Proposing a unification for the three complex forces, in addition to the potential properties and operation of this unification	183
Applications of introducing the "In sign" to the Schrodinger Equation	184
Identity stem waves ISWs and the idea of energy in shape gives a specific function	184

The potential collective operations of the unified imaginary complex forces.	185
The mechanism of the HD-force's entanglement	186
The notebook thought experiment	187
The spectrum of energies of the $HDBR_3$	187
The paradigm-shifting operation of the HD-force	187
The balancing force and the fine-tuning	188
Bibliography and references	190
Index	198

Introduction:

The Big Bang theory predicts that all matter, time, and space began at an incredibly tiny compact state about 13.82 ± 0.05 Billion years ago. This initial state was a hot, dense, uniform "soup" of particles that filled space uniformly and was expanding rapidly.

The scientist's observations of the movements of galaxies tell us that the universe is expanding at an accelerated rate.

We have three main possible scenarios for how our universe is expected to end: The big freeze, the big crunch & the big rip. The answer mainly depends on whether the force of gravity will win over the repulsive force expanding the universe, also known as dark energy, or the other way around. We are not sure yet which scenario will happen. The big freeze is expected to occur if the repulsive force wins over gravity. In that case, the expansion of the universe will not continue accelerating. Still, the universe will keep getting bigger, galaxies would separate, stars and planets would move away from each other. The light they emit is expected to be red-shifted to long wave lengths with very low energies. The universe is expected to become darker and colder, approaching a frozen state known as the heat death of the universe.

The big crunch is expected to happen if the force of gravity wins over the repulsive force. In that case, reversing the universe's expansion, galaxies are expected to start rushing towards each other, leading to a stronger gravitational pull. Space would get tighter, and temperatures are expected to rise, leading to an extremely dense, hot, and compact universe comparable to the state that preceded the big bang.

One of the proposals introduced in this book is that our universe is expected to end the same way it started. We are not sure whether it will be the big freeze, the big rip, or the big crunch. The "Big Defreeze" idea -proposed in this book as the scenario that happened at the beginning of the universe preceding cosmic inflation is compatible –at least partially- with any of these scenarios, which is explained in further detail later in this book.

The book suggests a potential mechanism that could be referred to as "The Big Defreeze," which could potentially explain what preceded inflation and the Big Bang. It could as well potentially explain the actual Bang of the Big Bang.

Additionally, the book goes a step before the proposed "Big Defreeze" to use an analogy called "The straight-line analogy," which could as well explain the driver of the "Big Defreeze" mechanism.

The book proposes another analogy titled "The straight-line analogy," which considers that our universe or even multiverse has a one-dimensional representation of reality in the form of a straight line. This analogy is based on the holographic principle, the cosmic inflation, the landscape of string theory, the topological manifolds and manifolds bundles.

Suppose you consider the two-dimensional data and the three-dimensional data as two versions of reality or two reconstructions of the same reality. Then why do we not have a third version or the third reconstruction of reality in one dimension represented in a straight line?

The analogy goes that we have a straight line as a one-dimensional representation of reality, and a tiny fluctuation happened on that straight line without affecting its geometry. This fluctuation was then projected on another space forming the identity stem wave of the universe ISWU. Fluctuations and interactions started happening to form an ISW crystal. The book proposes that when all matter –even if it is not bosons or ideal gases- is shattered to their elementary particles form, Bose-Einstein condensates BECs. These matter waves started fluctuating, forming what is referred to as the identity stem waves until the winding frequency of Helium's ISW causes symmetry breaking of the ISW crystal leading to the Big Bang and inflation. The book also addresses what could be referred to as "The Fermions Problem."

The book suggests as well that "The Big Defreeze" would require three imaginary "Complex forces":

- The R-force "The retrieving force."
- The B-force "The balancing force."
- The HD-force "The higher dimension force."

The book explains the definitions of these three forces, their properties, their potential operations, and how they could potentially work with physics's current known fundamental forces.

The book then proposes the "two-worlds interpretation." It briefly presents the many-worlds interpretation and the measurement problem. Then it explains the idea of the two worlds interpretation, which is simply that we have two worlds, one in our universe on a two-dimensional holographic film having quantum mechanical behaviors. And the other one is totally outside our universe or even the multiverse and its space-time parameters, which does not follow the quantum mechanical behavior.

The version of you in our universe is an observer with constituents following the quantum mechanical behavior, which requires the second version of you in the other world with constituents that do not have quantum mechanical behavior.

It is essential to mention that the book is built in the form of analogies and thought experiments. It is not claiming to be proposing scientific theories.

The proposals and ideas mentioned in this book will require further scientific validation and research.

Ever since the birth of modern science, abstract mathematical entities have played a significant role in formalizing physical concepts. Our current understanding of velocity was only possible through the introduction of derivatives. The modern conception of gravity owes to the invention of non-Euclidean geometry. Basic notions from representation theory made it possible to formalize the notion of a fundamental particle.

The new equations and notations introduced in this book are just indicative to explain the presented ideas. They might contain mistakes; they might require some complex numbers and Planck constant. They are just merely basic descriptions of the proposed ideas. Different notations or even a different mathematical framework could be used to describe these ideas.

Thought experiments and analogies have been one of the most successful approaches of humanity that led to some of the most significant discoveries in humankind's history. Throughout history, the greatest minds used thought experiments and analogies as a tool to understand the laws of the universe work, drive theories, laws, and equations.

Some of the most famous thought experiments are the ones that led to discovering and driving the laws of gravity, relativity, and quantum mechanics. The key has always been asking the right questions. It is not necessarily always the correct answer. But the right question is the key to opening more doors in science, philosophy, and any field of study.

Isaac Newton's famous question led to gravity "Why the apple fell on the ground while the moon did not?" The question seems quite basic and intuitive, yet a simple right question was asked, and a new era in physics started.

Einstein had always utilized thought experiments to understand and explain his revolutionary concepts as the famous train and lightning experiment for explaining special relativity. The accelerating elevator and person falling off the roof were thought experiments used to describe general relativity.

Schrödinger's cat is another thought experiment used to criticize the Copenhagen interpretation of quantum mechanics and the state of quantum superposition. Yet, it is often used in theoretical discussions related to quantum mechanics interpretations.

A background in physics would be a plus and is required for some sections in the book, yet you could still enjoy the book, the thought experiments, and the analogies without a background in physics.

If you are a physicist or a mathematician and at any part of the book, you felt that it makes no sense. I would recommend that you complete the book as potentially maybe one sentence at any section of the book could inspire you to develop something that makes some sense. Please make sure that if any part of this book inspired you to come up with something that makes some sense, to tell me first so we can write a paper together ☺.

Chapter 01:
The Big Defreeze

What could happen after the expected heat death of the universe?

As the universe is expected to become cooler and cooler, heading towards absolute zero, this book suggests that all matter, even if it is not a boson or ideal gas, is expected to form Bose-Einstein Condensates (BEC)[1]. (The fermions problem is addressed in this book as well).

When a gas is cooled down, the atoms slow down, and their energies decrease. Due to the quantum nature of atoms, they behave as waves that increase in size as the temperature decreases. At very low temperatures, the size of the waves becomes larger than the average distance between two atoms. At this very low temperature, all of the bosons can be the same energy in the same quantum state. They all form a single collective quantum wave called a Bose-Einstein Condensate (BEC).

This thought experiment suggests that in the scenario of the Big Defreeze at the beginning of the universe, all matter started with a single collective quantum wave which could be referred to as the Identity Stem Wave of the Universe (ISWU). In other words, the universe was in a "frozen state" at almost absolute zero temperature and potentially shrunk in a very tiny collective wave.

The thought experiment then suggests that our universe started with a reverse mechanism to a modified big freeze scenario (explained in this book). It began as a tiny collective quantum wave, which could be referred to as the **"Identity Stem Wave of the Universe"** ISWU. ***This stem wave started gaining energy leading to fluctuations behaving like a string forming what could be referred to as the "Identity stem waves of matter," leading to a super symmetrical crystal with no well-defined boundaries.***

---

[1] Bose-Einstein Condensates (BEC) which is a state of matter typically formed when a gas of bosons at low densities is cooled to temperatures very close to absolute zero at which case quantum mechanical phenomena, particularly wave function interference, become apparent macroscopically.

The "Identity stem waves" of all elements started forming and interacting. This is so far pretty much what is generally expected anyways.

Let's go further with the thought experiment. If all matter -even if it is not a boson or ideal gas- got smashed and shattered to their elementary particles. For example, let's consider iron shuttered to its elementary particle, and these elementary particles or, more accurately, fluctuations of waves started cooling down to degrees close to absolute zero forming Bose-Einstein Condensates. This BEC or quantum wave would act as the "blueprint" or the "Identity stem wave" of this matter.

The frequencies of these identity stem waves ISWs are comparable to the temperature at which they form BEC. This thought experiment suggests using the Fourier transformation to test and measure the mechanism of forming the ISWs as spikes/fluctuations. It indicates that for the ISWs of all matter, the winding frequency is comparable to the ISW's frequency except for helium; the winding frequency is expected to be 299,789,012 times the frequency of Helium's ISW, which suggests a symmetry-breaking mechanism. **This mechanism breaks the symmetry of the super symmetrical crystal leading to an explosion or a big bang with a value comparable to the speed of light.**

In simpler words, this thought experiment suggests that our universe was frozen, and it got heated to a degree leading to the big bang, and it is heading towards getting frozen again.

This proposed mechanism could help us understand the time frame of forming identity stem waves of all elements. These identity stem waves can lead to what can be referred to as the **"Big Defreeze Grand Spectrum,"** from where we can determine the identity stem waves of all elementary particles and how they were "born."

A suggested methodology for experimental validation would be: Shuttering particles of two periodic table elements in the cold atom lab and observing the reactions, including the potential formation of two identity stem waves and how they react. Please refer to the "Testability of ideas introduced in this book" section.

Understanding the "Identity stem waves ISWs" of matter, basically how matter was born, how the ISWs of different matter interacted with each other, and the time frame for that could help us create hybrid matter.

A disclaimer is mentioned in the introduction of the book, but I should mention it here again. This is just a thought experiment that requires further scientific validation and research.

The Big Defreeze compatibility

The Big Freeze is compatible with inflation and the Big Bang. With a tiny difference which is that the very first phase of the universe, which was tiny and condense, was not hot; it was cold. Then it got heated to a very high temperature through the symmetry-breaking mechanism of Helium's ISW. For more details, please refer to the "Testability of the ideas introduced in this book" section, including the Big Defreeze in the Cosmic Microwave Background part.

One of the proposals of the Big Defreeze is that the universe is going to end the same way it started. So if we know enough accurate information about how the universe is going to end, we will have good information about how the universe started and vice versa.

So how is our universe going to end?

Cosmologists have three possible main answers to this question.

- The big freeze.
- The big rip.
- The big crunch.

We will briefly present them here and then show how the idea of the big defreeze is sort of compatible –at least partially- with any of the three scenarios.

So, we have the attractive force of gravity and the repulsive force expanding the universe, also known as dark energy. And we know that the expansion of the universe is accelerating.

The big freeze is what happens if the repulsive force pulling the objects apart is just strong enough so that the expansion of the universe would not be able to accelerate anymore. But the universe would keep getting bigger. Clusters of galaxies would separate. The objects within the galaxies like suns, planets, and solar systems would move away from each other until galaxies dissolve into lonely objects floating separately in vast space. The light they emit would be red-shifted to long wavelengths, with very low faint energies, and the gas emanating from them would be too thin to create new stars. The universe would become darker and colder. Approaching a frozen state is also known as the big chill or the heat death of the universe.

The big rip is what happens if the repulsive force was so strong and the expansion of the universe continues to accelerate. It will eventually overcome the gravitational force tearing apart galaxies and solar systems and overcome the electromagnetic, weak, and strong nuclear forces that hold atoms and nuclei together. As a result, the matter that makes up the stars breaks into tiny pieces. Even atoms and subatomic particles will be destroyed.

The big crunch is what happens if the force of gravity brings the universe's expansion to a halt and then reverses it. Galaxies would start rushing towards each other, and as they clumped together, their gravitational pull would get stronger. Stars, too, would hurtle together and collide. Temperatures would rise as space would get tighter. The size of the universe would plummet until everything is compressed into such a small space that even atoms and subatomic particles would have to crunch together. The result would be an incredibly dense, hot, compact universe—a lot like the state that preceded the Big Bang.

Could this tiny point of matter explode into another Big Bang? Could the universe expand and contract over and over again, repeating its entire history? The theory describing such a universe is known as the Big Bounce.

If we follow the logic that suggests that the universe will end the same way it started, as proposed in this book. Then what would be the most likely scenario?

The intuitive answer would be the big crunch, as the universe would be back to a dense, hot, and compact state, a lot like the state that preceded the Big Bang. Yet, our most recent observations show that the expansion of the universe is accelerating. This means that the big defreeze or the big rip is more likely to happen depending on the strength of the repulsive force or the amount of dark energy.

We are looking here at a sort of counter-intuitive possible answer. Yet, this answer would be somehow -at least- partially compatible with some of the concepts of the three scenarios and our most recent observations.

Our most recent observations show that the expansion of the universe is accelerating, and in case the repulsive force was so strong, we will head towards a big rip where the matter that makes up the stars breaks into tiny pieces. Even atoms and subatomic particles will be destroyed. But what do we mean by destroyed? What are these particles fundamentally?

If you ask a group of particle physicists what a particle is, you might get different answers on the definition of the particle. One of the answers you might get is that "a particle is a collapsed wave function." Another explanation could be "a particle is a quantum excitation of a field." A third answer could be "What we think of as elementary particles, instead, they might be vibrating strings." (Please refer to the "What is a particle?" section in this book).

Whether we consider the particle as a collapsed wave function or a quantum excitation of a field or a vibrating string, the idea of the Big Defreeze suggests that when all matter is shattered to their elementary particles, and we start heading toward the heat death, in this case, all "shattered matter" even if it is not a boson or ideal gas are expected to form Bose-Einstein Condensates. The atoms will slow down, and their energies decrease. Due to the quantum nature of atoms, they behave as waves that increase in size as the temperature decreases. They all form a single collective quantum wave, a Bose-Einstein Condensate (BEC), which could be referred to as the ISW "Identity Stem Wave" of the universe "ISWU."

The identity stem wave of the universe ISWU would then start gaining energy leading to fluctuations behaving in the manner of a string, which would form what could be referred to as the "Identity stem waves of matter," leading to a super symmetrical crystal with no well-defined boundaries, where the "Identity stem waves" of all elements started forming and interacting.

So basically, this is greatly compatible with what the big rip suggests, where the matter that makes up the stars breaks into tiny pieces. Even atoms and subatomic particles will be destroyed. The Big Defreeze suggests that the shattered/destroyed atoms and subatomic particles form the BECs, and the identity stem waves ISWs.

The Big Defreeze is compatible as well with the Big Freeze in the part related to the heat death of the universe. There is a bit of a discrepancy, though, in the part where clusters of galaxies would separate, the objects within the galaxies like suns, planets, and solar systems would move away from each other until galaxies dissolve into lonely objects floating separately in vast space.

The Big Defreeze suggests that galaxies and objects within them would be shattered to their elementary particles due to the big rip and then frozen to their BEC/ISW state due to the heat death of the universe. And not in the form of frozen lonely, separated galaxies.

Probably the Big Crunch is the least compatible with the Big Defreeze. Yet the idea that the universe would be compressed into such a small space like the state that preceded the Big Bang is what the Big Defreeze is suggesting. The discrepancy is that this small and compact state was not initially hot and will not be eventually hot; rather, it would be a small compact frozen BEC/ISW state due to the heat death of the universe. Then due to quantum fluctuations, it will start gaining energy/heat and becoming hot.

In other words, the Big Defreeze suggests that the state that preceded the Big Bang was an initially small, compact, and frozen state, then it started gaining heat and energy.

What is the reason for these fluctuations? How did this BEC/ISW state gain energy/heat? This will be proposed in the straight-line analogy later in this book.

The idea of the Big Defreeze is pretty compatible as well with the Big Bounce. The possibility is there that this tiny point of identity stem wave explodes into another Big Bang and then follows the same scenario of shattering the newly formed matter to its elementary particles, which then get frozen to their BEC/ISW state. Yet the notion of the Big Bounce that there is no way to tell how many bounces could have already happened or how many might happen in the future, and that each bounce would wipe away any records of the universe's previous history would not necessarily be the case with the Big Defreeze idea.

As for the Big Defreeze, the idea of the Identity Stem Wave of the Universe "ISWU" would have the information of the frozen universe. It would be a very tough challenge to sort of decode this information. Yet basically, if we figured out the equations of how these identity stem waves evolve every time it bounces, there is a chance that we might know at least some information about the previous and potentially the future universes.

The Big Defreeze and the Bose-Einstein Condensates

Given that the primary mechanism of forming the identity stem waves ISWs is forming Bose-Einstein Condensates let's explore the story of the Bose-Einstein Condensates and see how to relate them to the idea of the Big Defreeze and the ISWs.

The Bose-Einstein Condensates is the coldest matter in the universe. And as the Nobel Laureate Wolfgang Ketterle -who won the prize in 2001 for the achievement of Bose-Einstein condensation in dilute gases of alkali atoms, and early fundamentals studies of the properties of the condensates- says **you can regard the Bose-Einstein condensates as matter made of matter waves.**

The identity stem waves ISWs of matter proposed in this book could be considered as a type of Bose-Einstein condensates. They are basically matter waves. There is nothing new in this concept, except potentially just the name to differentiate it. However, the idea of the Big Defreeze put it in the context that these matter waves were formed for all matter at the beginning of the universe before the Big Bang and inflation.

The Bose-Einstein condensates – as per Wolfgang Ketterle – is a form of matter where the quantum mechanical nature, the wave nature of matter manifests itself at a microscopic scale.

From the early days of quantum mechanics, we know that a particle that is classically characterized by position and velocity should rather be described as a quantum mechanical wave, a De Broglie wave. And the wavelength of the De Broglie wave (the quantum mechanical wave) follows the De Broglie relation. It's inversely proportional to the velocity of the particle $= \frac{h}{mV}$.

When you start learning physics, you usually struggle with the concept that all objects, you, me, microscopic and macroscopic objects, are waves and particles at the same time (Please refer to "what is a particle?" section in this book). Particles propagate as waves, but when we detect them, it makes a click in a detector.

So, if you and me, if we are all waves, why do we not perceive the wave nature of matter in everyday life?

How do we perceive if something is a wave or not in the first place?

If we want to explain why sound is a wave, a very reasonable answer would be: if two people talk to each other. They can hear each other even if they do not see each other because the wave propagates, deflects around corners.

What happens in the case of light?

It is known that light is an electromagnetic wave. It is not so obvious, though, because the wave length of light is so short that light is just deflecting a little bit, and you cannot perceive the wave nature of light very easily.

Now in the case of matter waves, the wave lengths are even shorter, they are given by the De Broglie relation, and **the velocity of a particle is determined by its temperature.**

The Big Defreeze predicts that the velocity of the identity stem wave of helium ISW_{He} was comparable to the speed of light. It caused the symmetry breaking of the ISW crystal, as explained earlier in this book. So in a sense, the velocity of the identity stem wave of helium was determined by its temperature.

The colder the temperature, the longer is the De Broglie wave length. Now talking about atoms at room temperature, the matter wave lengths (The De Broglie wave lengths) are smaller than an atom's size. So, therefore, the wave nature of atoms is not manifest. But now, we can cool down to the micro and Nano kelvin temperatures, and the De Broglie wave length becomes longer and longer.

These long De Broglie wave lengths are at the heart of the phenomenon of the Bose-Einstein Condensation.

What happens if you take gas in a container and cool it down?

The particles slow down. We should not describe the particles as sort of billiard balls; they are wave packets. They are quantum mechanical waves. But as long as the wave lengths are very short, those wave packets are localized, and we can follow their motion as if they were particles. And when we reach the point where the wave packets, the De Broglie waves, overlap, we cannot follow individual particles anymore. They become sort of a quantum soup of wave packets. At this point, the De Broglie wave lengths are comparable to the spacing of particles that a new form of matter forms. And this is the Bose-Einstein condensate.

In other words, if the De Broglie wave lengths become longer and longer and those matter waves overlap, then all the particles in the gas start to oscillate. And what they form is one matter-wave. This is what the Big Defreeze concept is based on. It predicts that the matter waves of all matter in the universe formed a collective matter-wave at the beginning of the universe. These matter waves started interacting and overlapping, forming a symmetrical crystal until the matter-wave of helium broke the symmetry of this crystal.

The difference between atoms in random motion and the Bose-Einstein condensate is analogous to the difference between the light from a light bulb and the light coming out of a laser.

The temperature fluctuations observed in the Cosmic Microwave Background $\Delta T/T = 10^{-5}$. These tiny fluctuations can be mapped.

When you analyze the properties of these fluctuations, you see that – besides all the local dynamics like sound waves, how the sound waves interact with the plasma, how the plasma interacts with the photons, etc. – overall, you find a very remarkable structure. You find that the waves that caused these tiny fluctuations in the temperature were generated with having the same power on all wavelengths. So they have what is referred to as the scale-invariant power spectrum. And they always come with an initial coherent phase condition as if they are coming from a broadband laser.

The Big Defreeze introduced in this book predicts that the source of this coherent, scale-invariant fluctuation is the identity stem wave of iron ISW_{Fe}. So, when the symmetry was broken by the identity stem wave of Helium ISW_{He}. The ISW of iron maintained its power on all wavelengths as it did during the ISW crystal phase in maintaining its power on all the other ISWs leading to the interactions in the ISW crystal. This prediction implies that the identity stem wave of iron ISW_{Fe} acted as the source of the laser beam-like fluctuations that we observe in the Cosmic Microwave Background.

If you want to achieve Bose-Einstein Condensation, you have to cool down a gas for Micro and Nano Kelvin temperatures until the De Broglie wave lengths are comparable to the spacing between atoms. The matter waves overlap and start to oscillate. The De Broglie wavelength depends on temperature; the distance between atoms depends on density. So the transition to Bose-Einstein Condensation is characterized by a combination of temperature and density.

Wolfgang Ketterle described watching how a condensate formed for him, like "how nature is giving birth to something very fragile, but we were able to observe it in its natural environment."

Laser cooling

The principles of laser cooling are pretty simple, you shine laser light on atoms, the atoms scatter light, and if you play some tricks, the light which is scattered/emitted has a shorter wavelength, is more energetic than the absorbed light. The scattering of light removes energy from the system, and the system cools down.

Magnetic traps

There are two types of magnetic traps. Some have a pointy potential, a linear, V-shape potential. And others have a round potential at the bottom; those traps are more tightly confining.

Interference between two condensates

Interference between two condensates allows you to photograph matter waves directly.

The interference is clear evidence for the wave nature. Wolfgang Ketterle and his colleagues needed two condensates and a laser knife cutting a condensate into two pieces. Then they switched off their magnetic container; the condensates fall due to gravity. They also expand because of zero-point motion and atomic repulsion. What they were hoping for is as the two condensates expand into each other and overlap to see the wave nature of those atoms. Then observe the interference patterns. They are interfering matter waves. So, the atoms spread out in the container. The faster the atoms are, the shorter is the De Broglie wavelength. So if you have a puff of gas and let it go, then take a snap shot. The atoms which are further out, have moved there faster and the atoms which are closer in are slower. So for the wave pattern of such a pulsed matter-wave, the matter wave length is longer where the atoms have not traveled so far and it is shorter where they have traveled further. They observed a distinct interference pattern. This interference pattern consists of perfectly straight lines.

And when they observe the overlap of the two condensates, they know that the Bose-Einstein Condensate is a coherent form of matter.

This was direct observation of the wave nature of matter. You take two puffs of atoms; they overlap, then you illuminate the atoms with a laser beam, and you take a shadow image.

A wave-like matter can propagate without any friction.

Vortices and Rotating condensates

Let's consider a rotating bucket experiment where we rotate an ordinary fluid, for example, a bucket full of water in one bucket. And spinning what could be referred to as a quantum fluid in the other bucket is a system where all the particles are just one wave.

If you rotate an ordinary fluid, upon rotation, it forms a parabolic surface. If you take, for example, superfluid helium and rotate it, the same happens. It is a parabolic surface. And it has to be like this because on a large scale; there is a correspondence equivalence between a classical system and a quantum system. But if you look closer at the quantum system, you will find that the system is littered with tiny little holes. Tiny little vortices, and around each vortex, we have full De Broglie wavelength. The rotating condensates have a Swiss-cheese-like appearance. The reason is the matter waves. We are not asking particles to go around and around; we are asking waves to go around. Due to the laws of quantum mechanics, we have a "snake" of waves. And the snake has to bite into its tail. It has to form a closed wave. So, therefore, we have the quantization condition that we need an integer number of the De Broglie wavelength around the circumference.

This structure of vortices is preserved upon the expansion of the gas. The faster you rotate a quantum liquid or gas, the more vortices you observe. We can study the properties of these vortices, their dynamics.

The Big Defreeze predicts that we had a quantum collective wave which could be referred to as the identity stem wave of the universe ISWU. Fluctuations and interactions started to happen in the ISW crystal forming the ISW of other matter, and the ISW of helium broke the symmetry of the ISW crystal.

It is expected that the ISWU would behave like a rotating Bose-Einstein condensate. It formed vortices; each vortex had a full De Broglie wavelength. By understanding these interactions and vortices better, their properties, and their dynamics, we would clearly understand what was going on inside the ISW crystal. If this idea was proven correct or at least partially correct, this would be a significant area of research.

The Rotation of the CMB in a BEC manner

It is expected as well that our universe is rotating in the sense that the whole CMB is rotating in a manner comparable to the rotation of condensate or a matter-wave, creating vortices and generating clusters that we see today in our universe. You can show that you find clusters or voids wherever you find deviations in the CMB from the mean temperature.

The "laser beam" of the identity stem wave of iron ISW_{Fe} would be sort of guiding these vortices. So, suppose we mimic and study the mathematics and dynamics of the vortices of the ISW crystal. In that case, this could be telling us something about the clusters formed due to the temperature variations that we see in our universe today. Yet, if we follow this line of thought, it would imply that the matter waves in our universe would continuously be updated given the nature of the new observers and measurements, which is different from the case in the ISW crystal at the beginning of the universe. The continuous update with information could be affecting the dynamics, energy, and even shape of the vortices.

The temperature fluctuations that we can observe in the Cosmic Microwave Background $\Delta T/T = 10^{-5}$. These tiny fluctuations can be mapped. When you look at the structure of these fluctuations, you can Fourier decompose them and then calculate the power spectrum of these fluctuations; you find that you can get evidence from how these tiny fluctuations form, that for instance, most of the gravitating matter in the universe is invisible, it is dark matter we only see it via its rotational effects among others on how these fluctuations were formed. It is also the stuff that keeps the galaxies together; if we do not have this dark matter, they will fly apart, and they will rotate too fast. And we also have evidence of dark energy which basically behaves like the cosmological constant, which has been now measured. It speeds up the expansion of the universe. It accelerates it.

If you take these tiny fluctuations, they cause temperature variations and also in the density of matter and therefore variations to the gravitational potential. You put it in a computer run it forward; you find that these tiny fluctuations generate all the large-scale structure of the galaxy clusters and superclusters that we see today in our universe. You can show that wherever you find deviations from the mean temperature, you find clusters or voids. You can map this.

When you Fourier decompose it and compute the power spectrum of the two-point function of these fluctuations, you see that there are distinct structures in there. There are waves, sound waves basically, that propagated through the hot plasma before it cooled enough that hydrogen could form.

When you analyze the properties of these fluctuations, you see that – besides all the local dynamics like sound waves, the way how the sound waves interact with the plasma, the way how the plasma interacts with the photons, etc. – overall, you find a very remarkable structure. You find that the waves that caused these tiny fluctuations in the temperature were generated with having the same power on all wavelengths. So they have what is referred to as the scale-invariant power spectrum. And they always come with an initial coherent phase condition as if they are coming from a broadband laser.

This could be a basis for the idea of the rotation of the CMB in a BEC manner forming vortices guided by the "laser beam" caused by the identity stem wave of iron ISW_{Fe}.

The synthetic material and optical lattice

Let's go back to the interference experiment by Ketterle and his colleagues. Where you build two condensates, so you basically have lots of atoms in the ground states in each of the two traps, and then you turn off the traps, let them go. Then you can measure the distribution of atoms as a function of position. They basically saw fringes in the atomic density because the two condensates were interfering with one another. What determines the spacing between the fringes? There is a De Broglie wavelength for the atoms, De Broglie wave length difference between the two condensates. So, you should estimate that under the conditions of this experiment, see fringes are separated by about 15 microns, which is what they saw.

So, this is quantum interference between atoms. But it is not like when you talk about electrons going one at a time through a pair of slits giving rise to an interference pattern as the double-slit experiment. In Ketterle's experiment, you see real-time interference between atoms like the real-time interference between classical electromagnetic waves. So it is the atom analog of a laser, an atom laser.

One of the things that have become a very active area is studying the effects of interactions among particles. And the ability to make BECs is sort of the standard tool used in those experiments nowadays.

One of the things you can do with BECs is that you can consider making a kind of synthetic material with an optical lattice.

An optical lattice is just an interference pattern set up between lasers. Say two counter-propagating lasers. You can control the intensity of the light and, therefore, the amplitude and the oscillations of the electric field in the standing wave. If you turn on such an optical lattice, starting with a Bose-Einstein condensate, turn it on slowly. This evolves to a kind of crystal that has one atom per potential minimum. It is like a kind of synthetic crystal. And unlike ordinary solid-state materials where you kind of stuck with the parameters and properties for the types of atoms in the crystal, in this case, we have the freedom to change the strength of the interactions between the atoms. In principle, they can interact by tunneling through the potential. An atom can tunnel through at some rate which depends on the width and the height of the potential. As we change the height, the tunneling rate changes. And so we can vary the rate at which they tunnel and study phase transitions that occur that depend on the relative strength of the on-site interactions when two atoms are sitting at the same side in the lattice and the strength of the tunneling from side to side. So what happens is for a given strength of the on-site interactions between atoms is that there is a kind of phase transition that occurs as we vary the height of the potential.

When the potential is very high, then the quantum ground state of that synthetic material is one where there is essentially one atom per site and no flow of atoms through the lattice; it behaves like an insulator. If we lower the height of the potential enough, then there is coherent tunneling from side to side, and it behaves like a conductor. The transition between the two is called the Mott transition.

Optical trap

The optical trap could be used as a transport mechanism for Bose-Einstein's condensates again, thanks to Wolfgang Ketterle and his colleagues. They now can form a condensate in one vacuum chamber, then focus the laser light on it and translate the focus by around 40 cms and carry the atoms in a new chamber. So they can deliver condensates with precision through pin holes and put condensates into new environments into micro traps and resonators, for example.

Hybrid matter

The reason for mentioning synthetic material, optical lattice, and optical trap here is to highlight the experimental approaches of such a concept; these experimental approaches could be applied to the identity stem waves of matter proposed in this book. These ISWs act as the very first DNA of matter from which matter sort of evolved later on. If this is proven correct or at least partially correct and we know enough information about these ISWs, we could be able to sort of creating hybrid matter. In the sense of working with the ISWs (analogically the very first DNA of matter) to harness properties and give birth to new hybrid matter. It is a way to go, but this could be one of the future research and applications of the idea of the identity stem waves ISWs.

The Fermions problem

Fermions and bosons are very different at the quantum level. As Wolfgang Pauli's famous exclusion principle states, identical fermions cannot occupy the same quantum state at the same time. Bosons, however, can share quantum states. But to observe this fundamental difference, gases of bosons or fermions have to be chilled to ultra-low temperatures, where individual quantum states have a high chance of being occupied. At these low temperatures, bosons will eagerly fall into a single quantum state to form a Bose-Einstein condensate, whereas fermions tend to fill energy states from the lowest up, with one particle per quantum state. At high temperatures, in contrast, bosons and fermions spread out over many states with, on average, much less than one atom per state.

Another difference is that fermions do not undergo a sudden phase transition in the ultra-low temperature regime. Instead, the quantum behavior emerges gradually as the fermion gas is cooled below the Fermi temperature $T_F = E_F/k_B$, where E_F is the Fermi energy – the energy of the highest filled state – and k_B is Boltzmann's constant. T_F, which is typically less than one μK for atomic gases, marks the crossover from the classical to the quantum regime.

A gas of atoms reaches quantum degeneracy when the matter waves of neighboring atoms overlap, i.e., when the thermal De Broglie wave length —which increases as the temperature falls- becomes about as large as the mean spacing between the atoms. The gas then can exhibit quantum behavior such as Bose-Einstein condensation (for bosons), Fermi pressure, and Pauli blocking (for fermions). At absolute zero gaseous boson atoms, all end up at the lowest energy state. Fermions, in contrast, fill the available states with one atom per state.

If the Big Defreeze is saying that all matter even if it is not, Bosons will form Bose-Einstein condensates when shattered to their elementary particles and cooled down to near absolute zero, Fermions might have a different opinion. We will have a "Fermions problem" here because their quantum behavior would be different.

The answer to that is that the Big Defreeze is dealing with the beginning of the universe. Basically, there were no bosons and fermions. This means that the quantum behavior of fermions is not to be considered at this stage. The Big Defreeze suggests that there was the collective identity stem wave of the universe ISWU. Fluctuations and interactions started to happen to lead to the ISWs of matter and the symmetry breaking by the identity stem wave of helium. All that does neither have fermions nor their quantum behavior.

The other question would be, but some of the predictions of the idea of the Big Defreeze is that the universe will end the same way it started. Won't we still have fermions towards the end of the universe? And the answer to that was partially explained earlier that in the case of the Big Rip, even the sub-atomic particles would be destroyed, they will lose their properties, including the quantum behavior of the fermions. But they are expected to maintain the behavior of forming BEC-like matter waves, which we refer to here as identity stem waves ISWs. So, after the sub-atomic particles are destroyed in the big rip scenario, the universe will be heading towards heat death and near absolute zero temperature, where the destroyed sub-atomic particles will start losing energies and forming BECs/ISWs, which will eventually shrink to again a collective identity stem wave of the universe ISWU.

The challenge is how do we test that now, as we still have fermions which maintain their quantum behavior?

The answer to this question could potentially be in the Fermion-to-Boson transformation; please check the bibliography for resources and citations.

According to a paper titled "The Fermion Boson transformation in fractional Quantum Hall systems," a Fermion-to-Boson transformation is accomplished by attaching to each Fermion a single flux quantum oriented opposite to the applied magnetic field. When the mean-field approximation is made in the Haldane spherical geometry, the Fermion angular momentum is replaced by $l_B = l_F - \frac{1}{2}(N-1)$. The set of allowed total angular momentum multiplets is identical in the two different pictures. The Fermion and Boson energy spectra in the presence of many-body interactions are identical if and only if the pseudopotential is "harmonic" in form. However, similar low-energy bands of states with Laughlin correlations occur in the two spectra if the interaction has a short-range. The transformation is used to clarify the relation between Boson and Fermion descriptions of the hierarchy of condensed fractional Quantum Hall States.

The authors stated that the sizes of the many-body Hilbert spaces for the Boson and Fermion systems were identical and that their numerical calculations verified that the mapping accurately transformed the ground state of the Fermion system into the ground state of the Boson system if and only if these ground states were incompressible FQH states. In the paper, they show that the F→B transformation leads to identical energy spectra if and only if the pseudopotential $V(L_{12})$ describing the interactions among the particles is of the "harmonic" form $V_H(L_{12}) = A + B L_{12}(L_{12}+1)$, where A and B are constants, and L_{12} is the total angular momentum of the interacting pair. Laughlin correlations occur when the actual pseudopotential $V(L_{12})$ rises more quickly with increasing L_{12} than $V_H(L_{12})$. Anharmonic effects (due to $\Delta V(L_{12}) = V(L_{12}) - V_H(L_{12})$) cause the interacting Fermion and interacting Boson spectra to differ for every value of the filling factor $v_F = v_B(1+v_B)^{-1}$. However, for appropriately chosen (short-range) model pseudopotentials, the F→B mapping accurately transforms the ground state of the Fermion system to that of the Boson system both for incompressible FQH states and for other low-lying states. The F→B mapping is also very useful in understanding the relationship between the Haldane hierarchy of Boson quasiparticle (QP) condensates and the CF hierarchy of Fermion QP condensed states.

Chapter 02:
The straight-line analogy

The frame of reference used to come up with this analogy are:

- The cosmic inflation.
- The landscape of string theory.
- The topological manifolds and manifold bundles.
- The holographic principle.

Let's start with cosmic inflation:

The conventional Big Bang theory does not say anything about what caused the expansion. It is a theory of the aftermath of the bang. In the scientific version of the Big Bang, the universe starts with everything already expanding with no explanation of how that expansion started. So the scientific version of the Big Bang theory is not really a theory of a bang. It is really a theory of the aftermath of a bang.

The conventional Big Bang theory says nothing about where all the matter came from. The theory assumes that for every particle that we see in the universe today, there was at the very beginning some precursor particle, if not the same particle, with no explanation of where all those particles came from.

In other words, the Big Bang theory says nothing about what banged, why it banged, or what happened before it banged. It really has no bang in the Big Bang. It's a bang-less theory despite its name.

Inflation fills in possible answers, very plausible answers for many of these questions.

What is cosmic inflation?

Cosmic inflation is basically a minor modification of the standard Big Bang theory. A good way to describe it is that it is a prequel to the conventional Big Bang theory. It's a short description of what happened before the Big Bang.

Most of the matter in the universe did form in the Big Bang; most of the matter is just hydrogen and helium. And about five different isotopes of hydrogen, helium, and lithium were primarily formed in the Big Bang. We can calculate the predicted abundances of those different isotopes, and the predictions agree well with the observation.

The combination of general relativity and modern particle theories indicates that at very high energy densities, there exist forms of matter that create a gravitational repulsion. In general relativity, gravitational repulsion is created by negative pressure. According to general relativity, it turns out that both pressures and energy densities can produce gravitational fields.

Inflation proposes that a patch of repulsive gravity material existed in the early universe. For inflation at the grand unified theory scale (~ 10^{16} GeV), the patch needs to be only as large as 10^{-28} cm. (Since any large patch is enlarged by inflation, the initial density or probability of such patches can be very low).

A plausible choice of when inflation might have happened would be when the energy scales of the universe were at the scale of grand unified theories. Grand unified theories unify the weak, strong, and electromagnetic interactions into a single unified interaction. That unification occurs at typical energy of about 10^{16} GeV (GeV is the energy equivalent to a mass of a proton). At those energies, we think that these states that create repulsive gravity are very likely to exist. And if that happened at that scale, the initial patch would only have to be 10^{-28} cm to ultimately lead to the creation of everything that we see on the vast scale we see it.

The gravitational repulsion created by this material was the driving force behind the big bang. The repulsion drove it into exponential expansion.

Small scale non-uniformity of the universe

The earth is about 10^{30} times denser than the average matter in the universe. The early universe –we believe- was uniform in its mass density to about one part in the 100,000 (10^5); still, there are some small non-uniformities. Things like the earth form because of these small uniformities, which come from quantum effects.

The landscape of string theory

There are maybe 10^{500} long-lived metastable states, any of which could serve as a substrate for a pocket universe. This is the landscape of string theory. How did the number 10^{500} come up? Please check the references.

Topological manifolds and manifold bundles

A bundle of topological manifolds is a triple (E, π , M).

E is the topological manifold called the total space.

π is the projection (surjective map) from the total space into the base space (It is a continuous map).

M is a topological manifold called the base space.

One of the examples is called the C-line bundle over some manifold M, which is a fiber bundle E $\xrightarrow{\pi}$ M with typical fiber C.

The C-line bundle over M plays an important role in quantum mechanics.

Let E $\xrightarrow{\pi}$ M be a bundle.

Then a map σ that starts in the base manifold M and maps into the total space (σ : M $\longrightarrow$ E), so it is not in the projection direction; it is the other direction. Such a map is called a section of the bundle if you can apply the map σ to go from M up to the total space, and then apply the projection π afterward, and if this leads you back to the identity on the base space id_M then you have a section. ($\pi \circ \sigma = id_M$).

In a special case, let's assume that the total space is the product of M & F (Fiber bundle).

E = M x F and that π is the projection to the first-factor π = $proj_1$

A bundle that is constructed like this is a product bundle (The Cartesian product of manifolds). Every product bundle is, in particular, a fiber bundle.

$\sigma : M \longrightarrow M \times F$

Only in this case, the cross-section can be understood as taking a point P in the base space and sending it to the pair (P, S(P)).

P $\longrightarrow$ (P, S(P))

Where s is a map from the base space into the typical fiber.

S: M $\longrightarrow$ F

If you tell me, for every point of the base space, the element of the fiber is just a specific function that you need to give me, what is the difference between the section and the function? Because from the function, I can construct the section. The answer is you only can if the total space is a product like M x F, but bundles are more general than products.

Physics example:

We could look at a C-line bundle over M in quantum mechanics. The C-line bundle is not necessarily the product of M with C; it is just a fiber bundle.

What is usually called a "wave function" in quantum mechanics is not a function; it is actually a section of the C-line bundle over physical space.

So, we have Product manifolds is contained in all the Fiber bundles, which are contained in all bundles.

Product manifold ∩ Fiber bundles ∩ Bundles.

The holographic principle

It is a tenet of string theories and a supposed property of quantum gravity that states that the description of a volume of space can be thought of as encoded on a lower-dimensional boundary to the region—such as a light-like boundary like a gravitational horizon.

A voxel is a three-dimensional version of a pixel. Can our world be described in both three-dimensional terms and two-dimensional terms?

Can you take three-dimensional information and re-express it? Voxels instead of pixels?

The answer is yes, but there is a high cost. And the high cost is when you take the three-dimensional data and try to lay it out in two dimensions, the result is always to scramble it horribly. It is always going to be incredibly mixed up. An example is a hologram.

A hologram film is a piece of the film incredibly scrambled; it's two-dimensional, with lots of scratchy little things. If you looked at it through a microscope, you would see no pattern, and you would not be able to tell what this thing was a hologram of.

Suppose you shine a light on the two-dimensional hologram film and an image form, a fully three-dimensional image.

Let's call it compressing data down to a two-dimensional surface but scrambling it beyond recognition in the process unless you know the detailed code.

The black hole horizon is a scrambled hologram of everything inside— two versions of reality, two reconstructions of the same reality.

The holographic principle: The maximum amount of information in a region of space is proportional to the area of the region.

The straight-line analogy

So, what if the multiverse, the landscape of string theory, and all the information within it have a one-dimensional representation of reality as well? If you consider the two-dimensional data and the three-dimensional data as two versions of reality or two reconstructions of the same reality. Then why do we not have a third version or, in other words, the third reconstruction of reality in one dimension represented in a straight line?

The logic would be the same as compressing three-dimensional data down to a two-dimensional surface; we will consider compressing two-dimensional data down to one dimension. The data would likely be even further scrambled beyond recognition in the process unless you know the detailed code. The code is embedded/scrambled in just one dimension, one straight line.

In the two-dimensional holographic film, it is required to shine a light on it to form a three-dimensional image. It could be as well the same logic to form two-dimensional construction from a one-dimensional construction of reality.

This light could be literally light waves/photons, it could be in the form of fluctuations/ripples, or it could be in the form of a force that could be added to our current known four forces. It could be as well one of the "imaginary complex forces" introduced in this book.

The holographic principle tells us that the maximum amount of information in a region of space is proportional to the area of the region. What if we want to know the maximum amount of information scrambled in a one-dimensional straight line. It would probably be proportional to its length.

How could we predict the length of the straight line?

Let's borrow the number from the landscape of string theory 10^{500} as this would be considered the maximum amount of information of a region in space. This region would be the straight line with all the information in the multiverse scrambled and encoded in it. That is one reason why the number 10^{500} from the landscape of string theory could be a good candidate for the length of the straight line.

Let's take the analogy further. We need to shine a light on the one-dimensional straight line to form a two-dimensional reconstruction of reality.

Let's consider we shine a tiny wave of light on our one-dimensional straight line. Let's consider the wave length of this light wave is $5*10^{-5}$, which will cause a tiny fluctuation that could be in the middle of the straight line or any other position on the straight line. This tiny fluctuation would be the first "Identity stem wave ISW" of the super symmetrical crystal explained in the first section of this book. This light and the resulting fluctuations will not be changing or affecting the geometry of the straight line. The fluctuations would be projected on another space. It will act as a tiny change in the information coded on the straight line which will be projected on another space.

If we followed the logic of this analogy, an inevitable question would come up, who shined this light and why?

There must be a force shining this light, and this force has to be continuous and unbreakable. Let's call this force the **HD force or the higher dimension force**, which will be presented later in this book.

Yet, this tiny fluctuation/ripple does not affect the straightness of the straight line; it might not impact it at all. Comparable to the way you shine a light on a two-dimensional holographic film. The three-dimensional images produced do not affect the holographic film.

This fluctuation does not lead to a wave in the straight line; it is a reconstruction of reality projected on a different space without affecting the properties of the total space or the straight line.

Let's consider this straight line a topological manifold called the total space comparable to the E topological manifold mentioned earlier. Then π would be the projection (surjective map) from the total space into the base space M. The mathematics of the topological manifolds and manifold bundles could be a good basis to understand this approach.

In a sense, it would actually be the total space having all the information scrambled in one dimension.

If you go back to the manifold bundles explained earlier. The wave function is a section of the C-line bundle over physical space. And from the function, we can construct the section.

A point could be projected from the total space or the straight line to the base space. Additionally, a section or a wave function, or a map can start in the base manifold M and map into the total space. Such a map is called a section of the bundle if you can apply the map σ to go from M up to the total space, and then you apply the projection π afterward, and if this leads you back to the identity on the base space id_M then you have a section. ($\pi \circ \sigma = id_M$).

To be clear about one of the significant thoughts this analogy is getting at -and without getting too philosophical- is that if we consider our universe, our world is the base space. Then simply our acts, our words, etc., which we basically emit in the form of waves, could be mapped in the total space and projected afterward on the base space. And we can measure it. We would be basically participating in coding the information on the one dimension scrambled straight line, which might be going through or having a two-dimensional reconstruction as well. In principle, the reconstruction could be in more than three dimensions as well.

Back to our small ripple with a wavelength of $5*10^{-5}$ (which is just a number for the analogy). In this phase, the universe is in a "heat death" state comparable to what will happen in the Big Freeze scenario. Actually, more accurately, at this phase, there was no heat yet. There was no energy yet. So, probably we should call it a "Pre-heat state" rather than a "heat death" state.

The Fourier transform will take it from here. The reason is that the Fourier transform would be our mathematical machine to decompose all the frequencies of information scrambled in that straight line and the super-symmetrical crystal. Additionally, if we worked it out the other way around, it could explain how this information was composed in the first place; at least for now, this will apply to the ISWs "The identity stem waves" of our super-symmetrical crystal. The Fourier transform would help us understand how these ISWs were composed and coded in the first place.

The first Identity stem wave ISW would be our first signal with a given frequency of 10^{-5} (It is just a number for the purpose of the analogy). At this phase, time and space are one thing, so we just represent them by a number 10^{-5}. Since we have a given frequency for our ISW, our mathematical machine will start treating it differently. Let's graph this frequency at a finite portion of space-time. Then take this graph and wrap it up around a circle. It could be considered as a rotating vector where the length of each point in space-time is equal to the height of that graph for that space-time which would be equal to 10^{-5} in our case here. High points of the graph correspond to a greater distance from the origin, and low points end up closer to the origin. So moving forward, 10^{-5} units of space-time would correspond to a single rotation around the circle.

So, here we have two different frequencies at play. The frequency of our identity stem wave ISW, which is 10^{-5}, and then separately, is the frequency with which we wrap the graph around the circle, which could be half of the rotation per space-time unit. The winding frequency at this phase matches the frequency of our signal 10^{-5}. All the high points on the graph happen on the right side of the circle, and all of the low points happen on the left.

Remember that the identity stem wave ISW is the BEC or quantum wave of the smashed and shattered elementary particle. So, as the first ISW starts to fluctuate, having a frequency and winding. This would create energy/heat and stimulate the creation/formation of other ISWs, which will have different frequencies and start winding.

Here a center of mass would start forming. As we add more frequencies and the graph winds up differently, that center of mass kind of wobbles around a bit. And for most of the winding frequencies, the peaks and the valleys are all spaced out around the circle in such a way that the center of mass stays pretty close to the origin.

It could get complicated, though when the sum of the winding frequencies is equal to the sum of the signal frequencies, the center of mass is expected to move far to the right. If we plot it on a graph, it will look like a spike. We will want to keep track of where the center of mass is for each winding frequency.

When a certain frequency persists for a long time, then the magnitude of the Fourier transform at that frequency is scaled up to more and more. The longer that signal persists, the larger the value of the Fourier transform at that frequency. For other frequencies, even if you increase it by a bit, this is canceled out by the fact that for longer time intervals, you are giving the wound-up graph more of a chance to balance itself around the circle.

The frequencies of these identity stem waves ISWs are comparable to the temperature at which they form BEC. This book suggests that for the ISWs of all matter, the winding frequency is comparable to the ISW`s frequency except for helium; the winding frequency is expected to be 299,789,012 times the frequency of Helium`s ISW, which suggests a symmetry-breaking mechanism. This is a mechanism that breaks the symmetry of the super symmetrical crystal leading to an explosion or a big bang with a value comparable to the speed of light. Remember that the velocity of the identity stem wave of helium is determined by its temperature, as explained earlier in this book.

The physics behind the analogy

From this paragraph until the end of the section, some of the technical backgrounds from physics that inspired the analogy will be presented, and some of the conclusions are relevant to the analogy. So, if you are not interested in physics and equations, you can skip to the next section.

Some of the mathematics of the Bose-Einstein Condensate, String theory, and the Higgs mechanism inspired this analogy, and some of them will be presented. It is pretty complicated physics, so there might be some glitches and mistakes that would require corrections. Please check the references section for the sources of the below-mentioned physics and equations.

In the case of an ideal gas, the onset of BEC occurs at inter particle spacing is comparable to thermal wave length. For the purpose of this analogy, let`s consider the same applies to the other elements of the periodic table when smashed and shattered to their elementary particles.

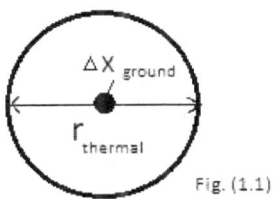

Fig. (1.1)

Where ΔX_{ground} shown in Fig. (1.1.) represents the ground state of iron`s identity stem wave ISW for example, let`s give at an approximate Tc value of 5.05×10^{-5} K (which is just a number as well for the purpose of the analogy). It also represents the width of the ground orbital.

Furthermore, $r_{thermal}$ represents the size of the symmetrical container or crystal -with no defined boundaries-where the identity stem wave of the iron, for example, is centered and symmetric.

The spread of the Gaussian wave function for a harmonic oscillator ground state is:

$$\Delta X_{ground} = \sqrt{\frac{\hbar}{mw_0}}$$

If we compare that to the thermal radius $r_{thermal}$ shown in Fig (1.1.) :

$$\Delta X_{ground} / r_{thermal} = \left(\frac{\hbar}{mw_0} \frac{mw_0^2}{T}\right)^{\frac{1}{2}} = \left(\frac{\hbar w_0}{T}\right)^{\frac{1}{2}} = N^{-1/6}$$

For T is comparable to T_E and Number of particles in excited state N_e (T_e) = the total number of particles N, and we set the total number of particles at this state to one, given that we only have one particle/wave at this stage, therefore :

$$\Delta X_{ground} / r_{thermal} = 5.05 \times 10^{-5} / r_{thermal} = (1)^{-1/6}$$

$$r_{thermal} = (5.05 \times 10^{-5})^{-6} = 5.05 \times 10^{-30}$$

ΔX_{ground} "identity stem wave" of iron was the first to form. It then acted as a black hole in a supersymmetric crystal absorbing and emitting radiations comparable to the hawking radiation and assisting in forming the "identity stem waves" of all the other elements of the periodic table. Elements that required smaller Tcs to form BECs or identity stem waves did not form except after interacting with the iron's BEC stem wave, which was the first to form.

How does this story fit into the Planck era?

Referring back to Fig (1.1.):

- The product of ΔX_{ground} and $r_{thermal}$ is comparable to the Planck's length, where we have a 10^{-5} and a 10^{-30}, which is comparable to 10^{-35} of the Planck's length as below the 10^{-35} and the identity stem wave of iron nothing will form or fluctuate.

- Let's consider the supersymmetric properties of the iron identity stem wave and the container or crystal. The $r_{thermal}$ is 10^{-30} the ΔX_{ground} is 10^{-5}; this suggests that at $10^{-2.5}$ fluctuations started happening, where $10^{-30} \times 10^{-2.5} = 10^{-32.5}$, which is comparable to the opposite of Planck's temperature. Less than this value, there will be no physical meaning for temperature.
- These identity stem waves could lead to what could be referred to as the "big defreeze grand spectrum" from where we can determine the identity stem waves of all elementary particles and how they were "born."
- It is worth mentioning that the 6.22404 in the estimated value of the Einstein temperature where all elements of the periodic table when shattered to their elementary particles form BECs is pretty close to the 6.62607015 of the Planck constant.

Let's consider a value of 6.22404×10^{-8} K represents the Einstein temperature where all elements of the periodic table, when shattered to their elementary particles, form BECs, and there are no fluctuations. This value could help us understand the time frame of forming the iron identity stem wave and subsequently the identity stem wave of the rest of the elements.

Where:

- At 6.22404×10^{-8}, we have the identity stem wave of the universe ISWU (the collective matter-wave) at the ground state with no energy and no fluctuations.
- At 5.05×10^{-5}, the first ripple/fluctuation started to happen, and the identity stem wave of iron ISW_{Fe} formed. This is expected to be at $10^{-32.5}$ units of time from the instant of forming the ISWU.
- The ISW_{Fe} started causing interactions, leading to energy/heat, which led to more fluctuations and forming of the other identity stem waves acting symmetrically until the identity stem wave of helium broke this symmetry with a velocity determined by its temperature and comparable to the speed of light.

- This could inspire as well the reason behind the value of the Planck constant at the beginning of the universe. As the minimum length for the ISWs would be in the order of approximately 10^{-35} or 10^{-34}, and the 6.22404 or 6.62607015 would be the coefficient of the ISWU derived from the 6.22404 $\times 10^{-8}$ value, which represents the Einstein temperature where all elements of the periodic table when shattered to their elementary particles form a collective matter-wave BEC, and there are no fluctuations yet.

This is, of course, speculative and requires further research and scientific validations. Yet, the line of thought leading to these numbers is suggestive of such potential conclusions.

The Higgs Mechanism:

The Higgs field gives mass to the gauge fields of a "spontaneously broken" gauge symmetry through the coupling to the Higgs, so it is the strength of the coupling between the Higgs and gauge field which determines in part the mass of the gauge field itself.

Breaking symmetry

$$L = \frac{1}{2}\partial_\mu \phi \partial^\mu \phi - v(\Phi) \quad (1)$$

For ϕ = Constant

$$\frac{\partial v}{\partial \phi} = 0$$

$$v(\phi) = -\frac{1}{2}\phi^2 + \frac{1}{4}\phi^4 \quad (2)$$

The symmetry of the potential is that:

$$v(-\phi) = v(\phi)$$

We will consider a new field ϕ (x) which uses a solution to the classical equation ϕ₀, and then we perturb small fluctuations about that solution ∂ϕ(x)

$$\phi(x) = \phi_0 + \delta\phi(x) \quad (3)$$

When we consider Equation 3 into the Lagrangian (Equation 1)

$$L(\phi(x)) = \frac{1}{2}\partial \mu(\phi_0 + \delta\phi)\delta^\mu(\phi_0 + \delta\Phi) \ldots\ldots$$

$$=$$

The derivative of the fluctuations – the potential v (ϕ) (Equation 2)

$$\frac{1}{2}\delta_\mu(\partial\phi)\delta^\mu(\delta\phi) + \frac{1}{2}(\phi_0 + \partial\phi)^2 - \frac{1}{4}(\phi_0 + \delta\phi)^4$$

For the solution ϕ = 0

$$L_0 = \frac{1}{2}\partial_\mu \partial\phi \partial^\mu \partial\phi + \frac{1}{2}\partial\phi^2 - \frac{1}{4}\delta\phi^4$$

This Lagrangian (the values of the equations of motion around φ) has the symmetry of ∂φ → - ∂φ

For the solution φ = 1

$$L_{+1} = \frac{1}{2}\delta_\mu \partial\phi \partial^\mu \phi + \frac{1}{2}(1 + \partial\phi)^2 - \frac{1}{4}(1 + \partial\phi)^4$$

This Lagrangian does not have the symmetry of ∂φ → - ∂φ

That is reflected in the fluctuations in those solutions.

If we consider the ISWs as fluctuations on a string, where we consider sigma as a line and x is the field, we apply the physics of the wave equation, the Lagrangian:

$$L = \frac{1}{2\pi}\int_0^\pi \left(\frac{\partial x}{\partial t}\right)^2 - \left(\frac{\partial x}{\partial \sigma}\right)^2 d\sigma$$

This brings us to the next chapter, "ISWs as strings."

Chapter 03:
ISWs as strings

The identity stem waves "ISWs" are massless. The massless particles can never be brought to rest even if you try to slow them down. They move with the speed of light. The Big Defreeze suggests that the reason they move with the speed of light is the symmetry-breaking mechanism by the identity stem wave of helium ISW_{He} which had a velocity determined by its temperature. For the ISWs of all matter, the winding frequency is comparable to the ISW`s frequency; except for helium, the winding frequency is expected to be 299,789,012 times the frequency of Helium`s ISW, which suggests a symmetry-breaking mechanism. The "freed" ISWs of all matter then will move with the speed of light following the identity stem wave of helium after the symmetry breaking.

Noether`s theorem tells us about conserved quantities and their connection with symmetries. For every symmetry, there is a conserved quantity, and in quantum mechanics, the conserved quantity could be called the generator of symmetry.

Supposing we have a Lagrangian and it depends on a bunch of coordinates (which we will consider here as the coordinates of the ISWs). We will call the coordinates qs. The Lagrangian $L = (q, \dot{q})$

So it depends on qs and the q dots (the time derivatives). The canonical momentum conjugate to a given q (related to the qth coordinate), partial of L with respect to the velocity:

$$p_i = \frac{\partial L}{\partial \dot{q}_i}$$

Suppose we have a symmetry that involves a transformation on the qs. It is infinitesimal symmetry where you just shift things by a little bit. So the shifting q under a particular symmetry operation, the ith number of q times some small parameter epsilon:

$$\delta q_i = f_i(q)\varepsilon$$

When you make a small transformation, each q changes by an amount that might depend on all the other qs times a small epsilon. In our example, the ISWs are the qs, and they transform by and the amount that depends on all the ISWs.

The conserved quantity could be, for example, angular momentum is constructed of ps and the variations of qs.

The generator of the transformation or the Noether charge is the sum of the ith momentum times the shift of the ith coordinate.

$$Q_{Noether} = \sum_i p_i f_i(q)$$

In quantum mechanics, $Q_{Noether}$ becomes the operator, which creates a small change. The operator action on a wave function induces symmetry.

It applies to our example where the $Q_{Noether}$ could operate on the wave function of the identity stem waves ISWs and induces the super symmetrical crystal of our ISWs.

ISWs could be treated as open strings. Two ISWs could operate in an open string-like behavior, they could potentially develop a soliton as well, and the physics of solitons could apply in this case.

Since the ISWs are massless at this stage, we will call the center of mass "center of vibration." As more ISWs center of vibrations joins, we get a free string direct from the center of vibration to the undefined "imaginary" boundary of the 2D super symmetrical crystal. On this string, the center of vibrations of other ISWs will be located; we will get a "loaded string" creating a field.

Hooke`s law tells us that the energy stored in a stretched string is proportional to the distance of stretching squared.

The low-lying spectrum of strings means the first few states. Let`s just remember that energy -in the formal analogy between non-relativistic physics and relativistic physics- becomes the square of the mass

$E \longrightarrow m^2$

So when we speak about the energy of a vibrating string and add them all up to find the total energy of the vibrating string corresponds to the square of the mass.

The discrete energy levels of the string -a string at rest (not moving in the perpendicular direction at all)- are a collection of harmonic oscillators. The first state and the lowest state is the state which is annihilated by all the annihilation operators. That is the ground state. All the annihilation operators, the As, and the Bs give zero.

$$a^-_n |0\rangle = 0$$

$$b^-_n |0\rangle = 0$$

The state in which none of these oscillators are at all excited. The 0 in $|0\rangle$ is not empty space; it is not vacuum. It is an unexcited oscillator. It is in the ground state of all of its oscillations. We would not say it has no oscillations; every harmonic oscillator has zero-point oscillations. Because of the uncertainty principle, there is some fluctuation. When we excite it, we will increase it by integer amounts of energy.

In order to excite it with the lowest energy excitation, we apply the creation operators, the lowest creation operator $a^+_1 |0\rangle$ each oscillator increase the energy by n units. That would be the oscillator with the smallest frequency which corresponds to n=1. The energy of that state is:

$$E = m_0^2 + 1$$

Another state which has one extra unit of energy is $b^+_1 |0\rangle$

If you take two eigenvectors of energy and you superpose them, it gives you back another eigenvector with the same energy. So any linear combination of $a^+_1 |0\rangle$ and $b^+_1 |0\rangle$ has the same energy.

What are the transformation properties of $a^+_1|0\rangle$ and $b^+_1|0\rangle$ when you rotate the space?

We have a preferred axis; we define the preferred axis. Boost the system up, so it is moving exponentially fast along the z-axis, but I still have a rotation symmetry of the entire thing about the axis that I boosted. $a^+_1|0\rangle$ and $b^+_1|0\rangle$ transform as vectors; they have vectorial characters. They have the same properties as the plane polarization of photons. The photon polarization is also a vector; namely, it corresponds to the direction of the electric field at some instant. It is a vector; it points in the direction perpendicular to the motion of the photon

For a string that has the properties of a circularly polarized particle

$$(a_1 \pm i\, b_1)|0\rangle$$

Its mathematics is identical to the mathematics of the polarization states of a particle moving down the axis. They transform into each other in the rotations. We could linearly superpose these with real coefficients, and then we would just be rotating the direction of oscillation, or we can add them with an i, and turn them into circularly polarized.

In string theory, the physics of a very fast-moving system as it moves in the perpendicular plane exactly has the form of two-dimension non-relativistic physics but what you have to keep track of is that the portion of the energy which is independent of the state of motion. The thing you would non-relativistically think of as the binding energy, just the energy at rest, is proportional to the square of the mass.

We would consider the identity stem waves ISWs as strings moving in two dimensions, and the binding energy could be the source of the retrieving force that will form after the two-dimensional plane turns into three dimensions. The proposed mechanism for the retrieving energy is mentioned later in this book.

Strings moving in two dimensions

For strings moving in two dimensions in a plane perpendicular to the direction of a large momentum, we want to make a model of a relativistic string which is wiggling around, moving, and stretching—but moving only in two dimensions, which is exactly comparable to the model of the identity stem waves ISWs.

We think of the string as a collection of mass points; in our ISW analogy, it could be the collection of the center of vibrations forming a field, as explained earlier. It is energy is the kinetic energy plus the potential energy:

$$E = \Sigma \frac{\mu \dot{x}_i^2}{2} + k \frac{\Delta x_i^2}{2}$$

Where k is the spring constant between neighboring mass points.

The string's Lagrangian would be the kinetic energy minus the potential energy.

We have N of mass points, and we want to take the limit in which N goes large. First, we need a parameter to label the mass points. Let's call this parameter σ (sigma), and it goes from 0 to π. It is just labeling.

We want the mass of the whole string to remain fixed, which means that the $\mu = \frac{1}{N}$ Where μ is a parameter that represents the non-relativistic analog mass.

The total analog mass of the system would be 1.

If we have a bunch of stiff springs and because of that, you cannot stretch it very much. If we combine in series a large number of these springs. The spring constant of the composite spring is $\frac{1}{N}$.

Let's consider:

$k = \frac{N}{\pi^2}$

$\Delta \sigma = \frac{\pi}{N}$

$$\frac{1}{N} = \frac{\Delta\sigma}{\pi}$$

Kinetic Energy: $\frac{1}{2\pi} \int_0^\pi \left(\frac{\partial x}{\partial \tau}\right)^2 d\sigma$

Potential Energy: $\frac{1}{2\pi} \int_0^\pi \left(\frac{\partial x}{\partial \tau}\right)^2 + \left(\frac{\partial x}{\partial \sigma}\right)^2 d\sigma$

If we thought of σ as a line in space and x as a field, this would be the energy of a simple wave field, and it would satisfy a wave equation. The physics here is exactly the same physics as all wave-like physics. The Lagrangian:

$$L = \frac{1}{2\pi} \int_0^\pi \left(\frac{\partial x}{\partial \tau}\right)^2 - \left(\frac{\partial x}{\partial \sigma}\right)^2 d\sigma + \frac{1}{2\pi} \int_0^\pi \left(\frac{\partial y}{\partial \tau}\right)^2 - \left(\frac{\partial x}{\partial y}\right)^2 d\sigma$$

Boundary conditions on the ends of strings

The conditions are actually determined by Newton`s laws because of this two-dimensional analogy. For the two-dimensional motion of a string, Newton`s laws apply.

Newton`s laws applied to the mass points (in our ISW analogy to the center of vibrations). It tells us how the mass points accelerate given the force on them. A typical mass point is being pulled from the left and the right, the forces won`t necessarily balance, but you will be getting a force from the left and a force from the right. There are two special points which are only getting forces from one side and not the other. That is the mass points at the end of the string.

What is the x-component of force at the endpoint of the string?

Hooke`s law tells that the force on the end of the string is the displacement or the distance between the N point and the N-1. In other words, the force on the Nth point of the string is proportional to $k\Delta x$. Where Δx is the separation between the Nth point and the N-1 point. And k is the spring constant.

$$N \frac{\Delta x}{\Delta \sigma} \Delta \sigma$$

$$\Delta \sigma = \frac{1}{N}$$

The force on the end of the string is proportional to $\frac{\partial x}{\partial \sigma}$, which is the amount that the string is stretched near the end. It's a stretching factor near the end.

By Newton's equation $\frac{\partial x}{\partial \sigma}$ has to equal the analog mass times the acceleration $\mu \ddot{x}$ which is equal $\frac{1}{N} \ddot{x}$

If we multiply $\frac{\partial x}{\partial \sigma}$ by N

$$N \frac{\partial x}{\partial \sigma} = \ddot{x}$$

We find that the acceleration will go to infinity as N gets larger and larger, so we impose Neumann boundary conditions in order to prevent infinite accelerations. In other words, in order to prevent the string from having infinite acceleration at the endpoint, we impose Neumann boundary conditions $\frac{\partial x}{\partial \sigma}$ is zero at the end points. When we compute the quantum mechanical energy levels, we get to know something about the masses.

Applying Neumann and Dirichlet boundary conditions to the identity stem waves ISWs

Imposing Neumann boundary conditions is basically to prevent infinite accelerations, in other words, in order to prevent the string from having infinite acceleration at the endpoint.

But could the mathematics be telling us something? Something that could be applied to our analogy of the identity stem waves ISWs, especially when dealing with the ISWs as strings.

We suggested earlier that for the ISWs of all matter the winding frequency is comparable to the ISW's frequency except for helium; the winding frequency is expected to be 299,789,012 times the frequency of Helium's ISW, which suggests a symmetry-breaking mechanism. **This is the mechanism that breaks the symmetry of the super symmetrical crystal leading to an explosion or a big bang with a value comparable to the speed of light.**

In this analogy, we will not impose Neumann boundary conditions to the identity stem wave ISW of helium which is behaving as an open string. We will let it accelerate to infinity which is another way of looking at an explosion or the Big Bang as an acceleration of an open string ISW breaking the symmetry of the two-dimensional super symmetrical crystal.

Except that the definition of infinity in this case in which the identity stem wave ISW of Helium is accelerating to is the Big Freeze or the heat death of the universe.

Following this line of thought, we can still apply boundary conditions, which would be Dirichlet boundary conditions on the ISW, to check if it will return to its initial state of BEC with the Big Freeze or the heat death of the universe. This might require applying the retrieving force or the R-force, which we will discuss later in this book.

The tension in strings & ISWs

In hadron physics, there is an idea that a hadron is a string that could be excited by setting it into rotation or setting it into vibration or by exciting the harmonic oscillators that make up the "stringy character" of the proton, for example. There is a certain energy scale. There is a certain amount of energy that each excitation will give you. In particular, the energy jump from the ground state to the first excited state. How much is it? What does it depend on?

In string theory, there is only one parameter. If you stretch a string from one point to another, it behaves pretty much like a spring, its energy (for a non-relativistic string):

$$E = k\frac{L^2}{2}$$

Where

k is a spring constant, L^2 is the square of the distance between the endpoints.

And that is the potential energy in a string. This non-relativistic energy is identified with the square of the mass on the string.

$$m^2 = k\frac{L^2}{2}$$

So, the mass of the string which is in the rest frame in which the string is at rest:

$$m = \frac{\sqrt{k}L}{\sqrt{2}}$$

The $\sqrt{k}$ has another name; it is called the tension. It is the tension in the string—the energy per unit length. The string is at rest; its energy is its mass. We stretch it out, the energy of it grows proportional to the length. It is like surface tension, except in this case, it is linear tension instead of surface tension. Energy is proportional to length; the coefficient is called tension. Tension in the string

If we work in favorite units in which the speed of light and Planck's constant are equal to 1 (c, h=1). In those units, energy has units of one over length $= \frac{1}{L}$. Given that, the units of the tension $\sqrt{k}$ would be energy squared or one over length squared. Let's call it T for tension, that tension in the string is the thing that sets the fundamental scale. It sets the scale for everything. It has units of energy squared $T = E^2$. It tells you that each excited state; in other words, each time you excite the string by one unit, it adds to its m^2 essentially that tension.

The bigger the string tension, the bigger the jump in mass when you excite something.

Let's take an ordinary spring and consider the mass of the spring to be one and k for the spring constant. The frequency is just square root of k

$$w = \sqrt{k}$$

And the energy E is also proportional to the frequency w.

So the energy jump between the ground state, the first excited state, the next excited state, and so forth are controlled by the string tension.

The string tension is energy per unit length. Force is also energy per unit length. Force times length is energy. Force times distance is work.

The tension is how much force is pulling you back if you pull the string apart.

$$E = TL$$

Energy = Tension * Length

Force is energy per unit length, so the weight the string can support is independent of its length. That is the character of these strings.

The string associated with fundamental particles like gravitons have a very large tension as predicted by string theorists, which means that the energy to excite a string is larger; these are very stiff strings. They have very large spring constants and very large frequencies of oscillation. Larger energies of oscillations require larger energies to excite them.

How much energy?

The thinking goes that it is probably somewhere near Planck energy.

The stiffer the strings are, the smaller they are. The larger the spring constant, the smaller the strings are. The thinking goes that the length of these strings is somewhere in the order of the Planck length/the Planck mass. How fast is one vibration? The Planck time. There is some reason to believe they are a little bit smaller than that, maybe a factor of ten or a factor of a hundred.

Applying the string tension equations to the ISWs

The string tension equations fit into our Identity stem wave ISW analogy. If we consider the ISWs as strings and initially there is no tension.

$$E = TL$$

Energy = Tension * Length

In that case, the tensions will not be zero (there is no zero value), simply the tension will just not be in the equation, and energy/mass would be the length of the string

$$E = L$$

Energy = Length

$$E = M = L$$

Energy = Mass = Length

This is a very fundamental concept for the identity stem wave ISW analogy. The mass or the energy is equal to the length of the ISW, which is treated as a string.

And as long as the frequency of oscillation of the ISW is equal to the winding frequency, this would probably apply. In the case of Helium's ISW, the winding frequency would be higher than the frequency of oscillation with a value comparable to the speed of light, which would lead to a very large tension.

Due to the mass-energy equivalence, the Planck constant also relates mass to frequency. A photon's energy is equal to its frequency multiplied by Planck's constant.

The idea that the mass or the energy is equal to the length of the ISW, which is considered as a string, also suggests a unified value for all forces at this phase. In other words, one value/number would have all the information required for the ISW at this phase. This value/number would represent the ISW's mass, energy, length, frequency of oscillation, etc.

So when we assume $5.05*10^{-5}$ for the iron's identity stem wave ISW_{Fe} this would represent the energy of the iron's identity stem wave ISW_{Fe}, its length (wavelength), frequency of oscillation, energy, mass (basically, it would be massless at this stage). In other words, $5.05*10^{-5}$ represents all the information required for the iron's identity stem wave ISW_{Fe}.

How do we drive it?

Evolution of information.

Evolution of information of the identity stem waves ISWs.

Schrödinger's equation is a very good candidate for understanding how the identity stem wave ISW's wave function evolves in time and what does it evolve to. How can we use a single number/value and then drive it with equations and constants to understand how does it evolve?

Some proposed notations and operations could help as well; please refer to the section in this book titled "Introducing the ln sign to the Schrödinger's equation." The analogy of the straight line and the complex forces, the $HDBR_3$ could suggest that these forces could affect the evolution of the identity stem waves ISW. Some proposed operations and notations are proposed in this book to approach the idea of the evolution of the identity stem waves ISWs.

Once the symmetry is broken, and the energy/momentum of the helium's ISW was sort of boosted along the axis perpendicular to the two-dimensional ISW crystal, we start dealing with more than one number/value; at this stage, we have:

- **Numbers/values** of the energy of the ISWs and the frequency of their oscillation. We can study the evolution of these numbers/values.
- **Proportions** before the helium's ISW broke the symmetry, the ratio of all the other ISWs frequency of oscillation to the winding frequency was one, so there was still no meaning for proportions. Yet once this ratio changes with the helium's ISW, we start defining the idea of proportions.
- **Angles** in three dimensions, where the helium's ISW break the symmetry and is boosted in a direction perpendicular to the ISW crystal.

In other words, the identity stem waves **ISWs now produce a system of numbers, proportions, and angles. This system becomes the basis of forming shapes and matter. These shapes are an expression of all energy principles within them**. This brings us to the equation introduced in my previous book, "Quantizing emotions – The Basic Mathematics of Psychology."

$$\hat{E} \ast \hat{s} = f\hat{\Psi}(x,t)$$

Where $\hat{E}$ is the energy operator, $\hat{s}$ is the shape operatee and $f\hat{\Psi}(x,t)$ is the resulting wave function. This proposed equation prescribes that energy in shape gives a specific function/wave function, which could be applied in principle to almost everything.

The energy is based on the total wave shape. The wave shape is expressed as the proportion of the wavelength to the amplitude. This can also be expressed by the angle that the wave crossed the central axis. In this context, the shape or the angle represents a function. The shape and the angle could be changed by introducing energy to the system, and this will not only result in a change in shape and angle but will lead to an update in the wave function, update of the information of the system. This also implies that we can change the information of the wave function/system by changing shapes/angles and by introducing new energies and information to it.

The resulting wave function or "IN product" should be considered an operator. In this proposed equation, we have the freedom to swap the in "IN product," the operator, and the operatee. In other words, the energy operator could act on the wave function to give a specific shape or topological manifold, etc. in that case; the equation would be:

$$\hat{E} \ast f\hat{\Psi}(x,t) = \hat{s}$$

Swapping the roles of the operator, operatee, and the "IN product" is a fundamental principle that could be useful in many conditions and various states.

This idea is not too exotic from superstring theory. T- duality is interchanging W (the winding number) and n (the momentum). In my proposed equation $\hat{E} \ast \hat{s} = f\hat{\Psi}(x,t)$, the energy operator $\hat{E}$ is comparable to momentum and the wave function is comparable to the winding number.

Can the shape of a drumhead be predicted from the sound that it makes, in other words, from the spectrum of vibrations? You can predict a lot of the shape of the drumhead. The same question applies to compactifications in string theory. From the spectrum of particles, in other words, the spectrum of vibrational energies, can you predict the shape and the size of the compact directions? The answer is yes to a large extent.

The question here goes beyond if we can predict the drumhead's shape from the sound it makes. The question is whether the spectrum of vibration can form the drumhead's shape? And whether both the shape and the vibrational energies can form certain functions or rather certain wave functions. How would that wave function evolve over time? And how would it affect and get affected by the surrounding ecosystem?

Back to our energy in shape gives a specific function equation

$$\hat{E} \divideontimes \hat{s} = f\hat{\Psi}(x,t)$$

The E here could as well be the Lagrangian of the standard model. The idea would be that we factor in shape/geometry. And what this is telling us is that when we introduce energy "in" shape/pattern, we get a function/wave function that evolves over time and could as well be tracked by Schrödinger's equation.

Could the idea of energy in shape gives a specific function $\hat{E} \divideontimes \hat{s} = f\hat{\Psi}(x,t)$ be extended to the idea of the patterns? Where the movement or rather the fluctuations of energy or complicated potential of energy in a particular pattern creates a specific function or a wave function, a proposed notation for this would be as follows:

$$\hat{E} \divideontimes \hat{P} = f\hat{\Psi}(x,t)$$

Where $\hat{P}$ Refers to the pattern in which the energy operator moves; following the same logic, the operator can affect the operatee and vice versa.

Similarly, the freedom to swap the "IN product," the operator, and the operatee. In other words, the energy operator could act on the wave function to give a certain pattern; in that case, the equation would be:

$$\hat{E} \divideontimes f\hat{\Psi}(x,t) = \hat{P}$$

An example of the geometrical pattern is what we can consider as the energy pattern in which the atoms are arranged in a molecule. This pattern is giving the molecule its function. The atom is formed of a specific pattern of electrons, and more fundamental "particles," the change in this pattern due to change in energy can change the function and lead to a different molecule. Let's go back to the idea that a particle could be considered a collapsed wave function. Could the collapse of the wave function target a specific pattern to stabilize and carry out the targeted function?

Back to $\hat{E} \divideontimes \hat{P} = f\hat{\Psi}(x,t)$. Even the energy operator in this equation might not affect & could be considered negligible in some situations.

This means that only the pattern can lead to a specific function. This analogous to the atom & molecule story. When atoms are joined in specific geometrical patterns, we get a particular substance's molecule. If the same types of atoms are joined in a different geometrical configuration or pattern, we get different kinds of molecules of other substances. In this case, the energy of the atoms played no role and the specific function/substance was formed just based on the geometrical pattern or configurations. In other words, the pattern led to a certain function, which could be expressed as

$$\hat{P} = f\hat{\Psi}(x,t).$$

The effect of the patterns on the resulting molecule could be directly measured and understood as we can study which pattern of atoms leads to which molecules and hence which types of substance; accordingly, we can understand which patterns lead to which sort of functions.

Every electron's location and energy in an atom is determined by a set of four quantum numbers that describe different atomic orbitals. An orbital is a region of probability where an electron can be found. These four quantum numbers are:

- "n" is the principal quantum number telling us the energy level.
- "ℓ" is the azimuthal quantum number that tells us the orbital type and angular momentum.
- "m_ℓ" is the magnetic quantum number, which tells us which specific orbital amongst the set and the projection of the angular momentum.
- "m_s" is the spin quantum number that tells us the spin. Each electron in an atom has a unique set of quantum numbers. The quantum numbers describe the pattern and the effect of this pattern.

If we considered two energy systems, for example, observer and observed energy system, are in resonance, exchange of information happens. Afterward, both will not return to their original state. They both keep part of the information, which causes change to both energy systems, and in a sense, they stay connected or entangled. This kind of entanglement is pretty much comparable to the act of measurement on a quantum system, where quantum entanglement happens with the measuring tool.

Riemann curvature tensor for the ISWs

The Riemann curvature tensor is a mathematical tool that helps us figure out of a space is curved or flat.

The abstract formula for the Riemann tensor:

$$R(\vec{u}, \vec{v})\vec{w} = \nabla_{\vec{u}}\nabla_{\vec{v}}\vec{w} - \nabla_{\vec{v}}\nabla_{\vec{u}}\vec{v} - \nabla_{[\vec{u}\cdot\vec{v}]}\vec{w}$$

The formula for the covariant derivative of a vector in a given direction:

$$\nabla_{\frac{\partial}{\partial x^i}}\vec{v} = \nabla_{\vec{e}_i}\vec{v} = \left(\frac{\partial v^k}{\partial x^i} + v^j \Gamma^k_{ij}\right)\vec{e}_k$$

The Γ^k_{ij} are the connection coefficients.

$$\Gamma^k_{ij} = \nabla_{\vec{e}_i}\vec{e}_i$$

There are many possible definitions for the connection coefficients. The most common coefficient is the Levi-Civita connection.

The main question we want to answer how can we tell if the identity stem wave ISW super symmetrical crystal -before the bang of the Helium ISW- is curved or flat? Did it become curved, and when? How did it transform from two dimensions to three dimensions?

What we need is a mathematical procedure for using the metric tensor to determine whether a space is curved or flat. We can also use the connection coefficients in this procedure because the connection coefficients are calculated using the derivatives of the metric tensor.

The tool which we will use is the Riemann curvature tensor $R^d{}_{abc}$ by detecting holonomy.

Holonomy is the twisting of a vector when transporting it around in a loop, and this is one way we can detect if a space is curved. If we have a flat space and we move a vector around in a loop, keeping it as straight as possible, the vector will be pointing in the same direction at the start of a loop and the end of the loop. However, if in some other space and we moved the vector around in a loop, and it is pointing in a different direction compared to the starting direction, the only possible explanation is that the space we are in is curved.

The thinking goes that we will consider all the ISWs as vectors, and we will get their covariant derivatives. It is expected that we will notice a different result for Helium's ISW, which will indicate that space started to change from flat to curved, and by calculating the tension tensor for Helium's ISW, it is expected to give us an indication of a symmetry-breaking mechanism.

We can potentially use the same mathematical tool using holonomic, geodesics, and parallel transport to check on the straight-line analogy.

The Higgs mechanism, ISWs, and spontaneous symmetry breaking

The Higgs field gives mass to the gauge field of a spontaneously broken gauge symmetry through the coupling to the Higgs.

The mechanism of things getting mass is tied to breaking symmetry; mass terms are nothing but interactions of a field with itself. Concavity determines the mass (the mass square). Goldstone massless bosons β are always going to appear anytime we have a continuous symmetry that we are breaking in this kind of Higgs mechanism; a continuous symmetry means that you have got a bunch of values that are connected along some flat direction, so if you fluctuate in that flat direction there will always be a massless mode associated with that. The time scale of the super symmetrical crystal could be referred to as the cradle crystal clock.

The application of this on our identity stem waves ISWs analogy is that they were massless fluctuations and started getting mass after the symmetry was broken by the Helium's ISW. This symmetry-breaking mechanism sort of set the 118 ISWs free, and they started traveling by the speed of light. Which is the speed the identity stem wave of helium broke the symmetry with.

A force that could be in action here is a retrieving force or the R-force which is working on retrieving the ISWs to their original state. The thinking goes that the R-force could be operating on the Helium's ISW to sort of decelerate it and bring it back to its initial state. But it is "too late"; things started to get mass, thanks to the symmetry-breaking mechanism. The R-force needs to find other ways and tools to achieve its goal. Gravity and dark energy (repulsive gravity) could be tools of the R-force to operate on the ISWs and later matter to retrieve them to their original state. The R-force could be considered as a "mother force" to gravity and dark energy.

This brings us to the idea of the imaginary complex mother forces.

Chapter 04:
The Imaginary Complex forces

On April 2021, the U.S. Department of Energy's Fermi National Accelerator Laboratory (Fermilab) announced the long-waited first results for the Muon g-2 experiment. The theoretical values for the muon g-factor are 2.00233183620(86), anomalous magnetic moment: 0.00116591810(43). The new experimental world-average results announced by the Muon g-2 collaboration are: g-factor: 2.00233184122(82), anomalous magnetic moment: 0.00116592061(41).

Immediately the results found their way to news headlines and the YouTube channels with titles like: "This result could change physics forever," "Have scientists discovered a new force?", "The Fermilab results strengthen the evidence for new physics."

To give a brief idea of these results and their significance. Until recently, all of the forces that we know of could be sorted into one of four kinds: Gravity, Electromagnetic, strong nuclear, and weak nuclear. These are known as the four fundamental forces of nature.

Are muons about to change that?

A muon is very similar to an electron, with the same charge, same spin, same properties, except for its mass of a muon is larger than the mass of an electron. Scientists have been studying the behavior of the muons when they are placed in a magnetic field, and they have been comparing the behavior that we expect to see based on the current physics that we know and the behavior that we do experimentally see. If our experiments show behavior quite far from what we expect from our theoretical understanding of physics, then perhaps the theory needs to be changed to reflect reality.

The muon is 200 times more massive than the electron. The probability of interaction between a particle and some massive virtual particle is proportional to mass squared. So, the muon is 40,000 times more likely than the electron to encounter a virtual Higgs boson, for example, or a virtual proton or other hadrons. It is 40,000 times more likely to encounter any unknown virtual particles.

So, what is going on here? Well, of course, this book does not have the answers. There are hundreds, if not thousands, of theoretical physicists who would be having plausible predictions and hypotheses.

Yet, the analogies used earlier in this book for the Big Defreeze, and the straight-line analogy might have some indications.

The Big Defreeze of the Universe is suggesting that our universe is expected to end the same way as it started. This would suggest that the big crunch is more likely to happen, except our most recent observations do not really indicate that the big crunch is the most likely scenario. Based on our most recent observations and the fact the expansion of the universe is accelerating, the big freeze or the big rip is more likely to happen. The big defreeze suggests that our universe started with a reverse mechanism to the big freeze with a small modification on the idea of the big freeze as explained in this book.

Following this line of thought, a force should be there to basically retrieve the universe back to the way/state it started with. This could end up being one of the functions of gravity, for example. But it could as well be a different force that could be referred to as **the R-force or the retrieving force** and could be considered as a "mother force" for gravity and dark energy.

Another inevitable force would be a **"Balancing force or the B-force,"** which would be necessary to sort of balance and maintain the geometry of the straight line, which is a one dimension construction of reality with all the information scrambled and coded into it. The B-force could be responsible for some of the constants of nature.

The third force that is likely to be there is the force causing the fluctuation/ripple on the straight line which we referred to earlier as the **"Higher dimension HD-force."**

So we have three complex forces:
1. The retrieving force (The R-force).
2. The Balancing force (The B-force).
3. The Higher dimension force (The HD-force).

Let's go through them one by one and explore their properties and the reason they could be considered as separate fundamental forces or at least imaginary complex forces.

We will propose some of their potential properties and operations. Additionally, we will propose a potential framework of mathematics and notations to describe these operations. Yet, these equations and notations would be just indicative and could have some mistakes that could be corrected, or we can use a different mathematical framework to describe the proposed operations of these imaginary complex forces.

The retrieving force (The R-force)

The retrieving force would basically be the force working on retrieving our universe to the state it started with. It would operate in a different way from gravity. It could be considered as a "mother force" for gravity and dark energy. The potential properties of the R-force are:

- One of the potential functions of the retrieving force could be the decay of the sub-atomic particles to another sub-atomic particle in the sense of phase transition to another state in an attempt to be retrieved to its original state.

- During the retrieving process, the particle/wave will undergo partial retrieving phases which could be referred to as partial retrieval or phase retrieval in forms of decay or changing states. During each retrieving phase, the particle/wave will gain information/experience that would affect its next phase or its reconstruction.

- The conditions/states in which the retrieved particle/wave ends up with or during the retrieving phases will not always necessarily be the same exact initial states/conditions.

- The retrieving force could have the ability to change time into space and vice versa. It could create and annihilate spaces acting on or using creation and annihilation operators.

- The retrieving force would operate on a particle beyond space and time boundaries with a unified space-time operator. This is counter-intuitive; we will use some analogies to try to approach the idea.

- The retrieving force would be creating and annihilating space & time.

- Gravity and dark energy could be tools for the R-force to retrieve the ISWs and matter to their original state.

The R-force could be responsible for the anti-matter using annihilation operators to annihilate matter.

We want to examine how our imaginary R-force could operate on the muon, for example, given the above-mentioned properties. But before we start. What is a particle in the first place?

What is a particle?

In the late nineteenth century and with the discovery of the electron, we knew that they are negatively charged "particles." This was based on the work of the scientists during the 1880s and `90s when they searched cathode rays for the carrier of the electrical properties in matter; their work culminated in the discovery of the electron in 1897 by the physicist J.J. Thomson. It only took us less than three decades to discover the electron-possessed spin. With progress being made in physics generally and quantum mechanics, specifically, our understanding of "particles," including electrons, started developing.

If you ask a group of particle physicists what a particle is, you might get different answers on the definition of the particle. One of the interesting definitions is that a particle is **"a collapsed wave function."** The wave function representing an electron, say, is spatially spread out so that the electron has possible locations rather than a definite one, but when you measure the electron`s location using a detector, its wave function collapses to a point, and the particle clicks at that position in the detector. Making a measurement is establishing an entanglement between the detector/device and a system.

Another interesting definition of the particle would be **"A particle is a quantum excitation of a field"** Quantum field theory is the math where particle physics is written. In that, there are a bunch of different fields, each field has different properties and excitations, and they are different depending on the properties, and those excitations we can think of as a particle. As physicists discovered more of nature's particles and their associated fields, a parallel perspective developed. The properties of these particles and fields appeared to follow numerical patterns. By extending these patterns, physicists were able to predict the existence of more particles. "Once you encode the patterns you observe into the mathematics, the mathematics is predictive; it tells you more things you might observe," explained Helen Quinn, emeritus particle physic at Stanford University. The patterns also suggested a more abstract and potentially deeper perspective on what particles actually are.

"What we think of as elementary particles, instead they might be vibrating strings" – Mary Gaillard. Physicists are trying to fit symmetries inside a single, larger group of transformations. The idea is that particles were representations of a single symmetry group at the beginning of the universe. Researchers placed high hopes in string theory: the idea that if you zoomed in enough on particles, you would see not points but one-dimensional vibrating strings. You would also see six extra spatial dimensions, which string theory says are curled up at every point in our familiar 4D space-time fabric. The small dimensions' geometry determines the properties of strings and, thus, the macroscopic world. "Internal" symmetries of particles, like the SU(3) operations that transform quarks` color, obtain physical meaning: These operations map, in the string picture, onto rotations in the small spatial dimensions, just as spin reflects rotations in the large dimensions. "Geometry gives you symmetry gives you particles, and all of this goes together," Nanopoulos said.

The theoretical physicist David Tong explained is that there are fields that underlie everything. And what we think of as particles are not really particles at all; they are waves of these fields tied up into little bundles of energy. There are no particles in the world; our universe's basic fundamental building blocks are these fluid-like substances that we call fields.

The Spin of massless particles

The spin of massless particles is a bit different in some ways than the spin of massive particles. The basic difference is that massless particles can never be brought to rest. They move with the speed of light; even if you try to slow them down, they still move with the speed of light. Whereas particles with mass, you can slow them down and bring them to rest, and that makes the difference in certain properties about spin.

The sleeping muon analogy

Let`s consider Alice is a muon. She will decay from Alice muon to Alice electron in about eighty years. The question is, during these eighty years, is there anything beyond the space and time Alice is recognizing affecting her, something that she can observe?

The answer is yes, while Alice is sleeping. The idea is that Alice does not feel the normal space and time while she is sleeping. She has no sense or at least a different sense of time and space. Time & space could be different; time could be space, and space could be time. Time and space could be one thing. We do not really have a scientific explanation for that. But as far as Alice is concerned, her sense of time and space are different. She is not in our world. Yet, the massive Alice is still physically sleeping on her bed. There are some physical, biological operations going on in her body within the boundaries of space and time of our world. Operations that she does not feel and does not control during his sleep.

But is that different space and time that we cannot scientifically explain yet affecting the massive Alice in our world within the boundaries of space and time of our world? The answer is we do not really know.

There are chances that what Alice is experiencing in this different space and time affects her on a very fundamental and subtle level that we do not really understand or are not yet able to track and measure.

Maybe we cannot yet measure its effect on Alice, but we have better chances of measuring it on an actual muon. This would be the retrieving force acting on muon. The retrieving force would be operating on the muon with space and time of different properties causing minor effects in the initial state of the muon that would probably trigger further changes leading to phase transitions in a butterfly effect manner.

The analogy might not be the most accurate description; it is not suggesting that dreams are a retrieving force operating on Alice. But, it is just an attempt to clarify the probability of a force that is different from gravity acting on the muon with different space and time properties and parameters.

Another inevitable question would come up. Is the retrieving force the only force beyond our space and time operation on Alice or our muon?

The analogies used in this book would suggest that there are two more forces beyond space and time that could be acting on Alice or the muon. These would be the balancing force and the higher dimension force. Both would qualify as separate imaginary complex forces as they have different properties and functions.

The potential mathematics of the R-force:

The potential mathematics of the R-force would be based on the mathematics mentioned in my book "Quantizing emotions – The basic mathematics of psychology," published on 17th of February 2021. It would be analogous and just an attempt to approach what could be the concept of the potential mathematics of the R-force. It could have glitches and mistakes, and it would require further research and scientific validation. A different mathematical framework could be used as well to describe the operations of the R-force.

Before we dive into the potential mathematics, let`s review what would be the basis of it from the book "Quantizing emotions – The basic mathematics of psychology." That would be the energy in shape = a wave function.

Introducing the "IN sign" to Schrödinger's equation

This part is based on my previous book, "Quantizing emotions – The basic mathematics of psychology." A disclaimer that I have to mention as a reminder, what I am proposing here is just an approach to the subject from a different perspective, and it will require scientific validation.

I will explain some technical background that led me to propose introducing the "IN sign" to Schrödinger's equation. The thought experiments presented in my previous book "Quantizing emotions – The basic mathematics of psychology" titled:

"The correct answer of one plus one," "The IN sign," "Quantizing emotions, "Quantum party," and the "Quantum ride" could help build the intuition and logic behind introducing the "IN sign" to Schrödinger's equation.

Max Planck quantized energies in multiplies of $\hbar\omega$ and then came Einstein said the energy of a photon is $\hbar\omega$ and the momentum of the photon is $\hbar\vec{k}$. So $(E,\vec{p}) = \hbar(\omega,\vec{k})$. Then came De Broglie with the idea that even though this was written for photons, it was valid for particles

as well, all particles and these particles are waves $\Psi(x,t) = e^{i(kx-wt)}$
Since k is positive, this is a wave moving to the right.

The operator that realizes the momentum would be $\hat{P} = \dfrac{\hbar}{i}\dfrac{\partial}{\partial x}$

The operator that realizes the energy would be $\hat{E} = i\hbar\dfrac{\partial}{\partial t}$

When the energy operator operates on the wave function $i\hbar\dfrac{\partial}{\partial t}\psi = \hbar w \psi$

$$i\hbar\dfrac{\partial}{\partial t}\psi(x,t) = E\psi(x,t)$$

This equation is prescribing how a wave function of energy E evolves over time. It tells you if you know the wave function and it has energy E, the left side of the equation is how it looks later, and you can take the derivative on the left side of the equation and solve this differential equation. In this equation, E is a number. If you know you have a particle with energy E, that is how it evolves in time.

So came Schrödinger and looked at this equation, which is true for any particle that has energy E. Maybe I do not know what the energy E is. One single replacement was introduced to this equation, replacing the energy E with the energy operator $\hat{E}$. Schrödinger's equation does not assume that the energy is a number because you do not know it. In general, if the particle is moving in a complicated potential, you do not know what the possible energies are.

$$i\hbar\dfrac{\partial}{\partial t}\psi(x,t) = \hat{E}\psi(x,t)$$

This is symbolically what must be happening because if this particle has a definite energy, then the energy operator $\hat{E}$ gives you the energy acting on the function.

Given that, what led me to introduce the "IN sign ✹" to Schrödinger's equation is if the particle (basically wave or, in other words, collapsed wave function) that the energy operator is operating on does not have definite energy? What if the energy operator is operating on a complicated potential or, in other words, a field of waves, which is fundamentally how energy operators operate. They do not operate on a single wave. They operate on a field of waves.

The energy operator requires the freedom to act on the operatee's multiple quantum states. Both the operator and the operatee can affect each other, so this requires the freedom to swap roles.

Additionally, we would require to include other operations being acted upon by the energy operator, not just position and time. The other operatee could eventually probably be reduced to position and time. However, we will need to define more operatees during the operation.

$$i\hbar \frac{\partial}{\partial t}\psi(x,t) = \hat{E} \ast \psi(x,t)$$

This equation would be the simplest form of introducing the in sign to Schrödinger`s equation. The energy operator is acting "in" a wave function with definite energy, so basically, the IN sign here is doing nothing.

$$i\hbar \frac{\partial}{\partial t}\psi(x,t) = \hat{E} \ast \hat{\psi}(x,t)$$

Here the equation is prescribing that the energy operator is acting "in" a wave function, which is the operatee in this case, and this wave function does not have definite energy; it could be a field of waves. This equation also prescribes that the energy operator and the wave operatee affect each other and can swap roles.

So far, we are trying to understand how the wave function evolves over time. What if we want to know how the wave function evolves relative to other factors, shape, for example? This shape could be itself evolving over time and could be time-independent; these are pre-defined conditions that we should define before carrying out the operation. These pre-defined conditions will help us conclude whether we will require the first or second derivative of the "IN product" to get the information we are looking for, which will take us to our next thought experiment, "Energy in shape = A wave function."

Energy in shape = A wave function.

This was based on the thought experiments presented in "Quantizing emotions – The basic mathematics of psychology" book titled: "The correct answer of one plus one," "The IN sign," "Quantizing emotions, "Quantum party," "Quantum ride," Introducing the "IN sign" to Schrödinger's equation.

We considered the energy as an operator on or rather "in" shape, which is the operatee. However, they can swap roles. The "IN product" would be a wave function that could be defined in terms of position and time or any other parameters.

$$\hat{E} \divideontimes \hat{s} = f\hat{\Psi}(x,t)$$

This equation is how it potentially look like. Where $\hat{E}$ is the energy operator, $\hat{s}$ is the shape operatee and $f\hat{\Psi}(x,t)$ is the resulting wave function. This proposed equation prescribes that energy in shape gives a wave function, which could be applied in principle to almost everything.

The concept of the shape as an operatee or even an operator in this proposed equation goes beyond the conventional shapes to topological spaces construction, construction of new topologies from given topologies, topological manifolds, and bundles, etc., which is a big area that could be explored. The idea is that shape/space acts as a container for energy. Energy can operate on the shape and change it, and vice versa shape can operate on energy and change its frequencies. The shape's concept goes even further to the idea of the pattern where energy can operate on a pattern and change it. The opposite could happen as well, where the pattern can operate on energy and change it. The pattern can cause a specific function independent of energy as well. Which is all explained in this book.

The resulting wave function or "IN product" should be considered an operator. In this proposed equation, we have the freedom to swap the in "IN product," the operator, and the operatee. In other words, the energy operator could act on the wave function to give a specific shape or topological manifold, etc. in that case; the equation would be:

$$\hat{E} \divideontimes f\hat{\Psi}(x,t) = \hat{s}$$

Swapping the roles of the operator, operatee, and the "IN product" is a fundamental principle that could be useful in many conditions and various states.

This idea is not too exotic from superstring theory. T- duality is interchanging W (the winding number) and n (the momentum). In my proposed equation $\hat{E} \divideontimes \hat{s} = f\hat{\Psi}(x,t)$, the energy operator $\hat{E}$ is comparable to momentum and the wave function is comparable to the winding number.

This could be extended to the idea of patterns as explained earlier in this book, where the movement or rather the fluctuations of energy or complicated potential of energy in a particular pattern creates a specific function or a wave function; it could as well written as follows :

$$\hat{E} \divideontimes \hat{P} = f\hat{\Psi}(x,t)$$

Where $\hat{P}$ refers to the pattern in which the energy operator moves; following the same logic, the operator can affect the operatee and vice versa.

How to relate this to the potential mathematics of the R-force?

The main function of the R-force would be retrieving the universe to its initial state. Here the function would be the main player here; it would create space, annihilate & create space. It will use energy in order to that.

So, potentially in order to use notations for this idea, we can use:

$$\hat{E} \divideontimes \hat{s} = f\hat{\Psi}(x,t)$$

We will add some notations

$$Ra_n^+ f\hat{\Psi}(x,t) = \hat{E} \divideontimes \hat{s}$$

Where:

- R refers to the retrieving force or the R-force
- a_n^+ Refers to a creation operator and if it could be a_n^- An annihilation operator.
- $f\hat{\Psi}(x,t)$ Refers to the wave function used by the R-force to whether operating on, create or annihilate energy and space.

The equation is a very basic notation telling us that the retrieving force, the R-force, is using a creation or annihilation operator through wave functions to operate on, create or annihilate energy and space. The second part of the equation $\hat{E} \divideontimes \hat{s}$ is dealing with the energy and shape formed, how they affect and interact with others as a result of the R-force's operation on them. And how they have the freedom to swap roles.

The Balancing force (The B-force)

The straight-line analogy suggests a construction of reality in one dimension of a straight line. This one-dimension construction of reality could precede the two and three dimensions construction of reality. Or they all could happen to exist at the same time, or they could have different space-time parameters.

This straight line is an energy configuration. Information is encoded and scrambled on this straight line, which would require balancing properties to maintain its straightness, geometry, compactness, and dimensionality properties. This would be achieved by a balancing force.

Some of the properties or characteristics of the balancing force would be:

- Repetitive.
- Continuous.
- Descriptive.
- Unlimited.
- Unbreakable.
- Reproducible.
- Duality.

The balancing force could be the main force behind some of the fundamental constants of our universe. The cosmological constant could be an example that will be presented shortly in this book.

The main mechanism of how the balancing force would operate is likely to be resonance.

Before we discuss the balancing force in further detail, let`s first talk about resonance for a little bit.

Resonance

Classical resonance

Resonance is something where we have some variable, and it varies periodically. The periodic variation usually requires that you drive the system. So, you first drive it, and then the system oscillates. You drive the system with a variable frequency then you observe a peak.

So, the phenomenon of resonance is that when you have something which can periodically vary and when you drive it, you see a peaked response when driven with a variable frequency.

Wolfgang Ketterle, the famous physicist and Nobel Laureate, mentioned a very interesting quote during one of his lectures at MIT:

"Resonance is the language we talk to atoms with."

Resonance at a certain frequency is finite damping. That would mean after the system is driven, the oscillation does not last for an infinite amount of time. And that implies that when we drive the system and look at the response as a function of frequency, there is a finite width for the driven system. And in many ways, the damping time and the finite width are related by Fourier transform.

Atomic physics is interested in every single possible aspect of this resonance, the shape of the curve, how it can be modified, what happens when we tie it strongly, what happens when we tie it weakly, etc.

We usually characterize oscillators by the sharpness of the resonance.

The sharpness of the resonance is the ratio of the beats of the resonance and the frequency or the inverse of it. So if you have an oscillator at a kilohertz and the resonance is one hertz wide, we say the resonance has a Q (A quality factor) of a thousand. And that means you can observe a thousand oscillations before the oscillation decays away.

The Q factor explains how fast energy decays in an oscillating system. It compares the frequency at which a system oscillates to the rate at which it dissipates its energy. The sharpness of resonance -described by the Q factor- increases or decreases with an increase or decrease in damping. As the amplitude increases, the sharpness of resonance (The Q factor) decreases. Higher Q indicates a lower rate of energy loss, and the oscillations die out more slowly. A pendulum suspended from a high-quality bearing, oscillating in air, has a high Q, while a pendulum immersed in oil has a low one. Resonators with high-quality factors have low damping so that they ring or vibrate longer.

The frequency of light is 10^{15} HZ. The wavelength is 300 nanometer.

For example, the Q of the earth's rotation fulfills all of our requirements for resonance and oscillations. It is a periodic phenomenon. It has a Q of 10^7. So, the precision of the rotation of the earth is better than one part in a million. The quality factor Q for a pulsar star is 10^{10}.

If a resonance has a high-quality factor, the more sensitive it will be too tiny little changes. A high-quality oscillator is a tool for discovery.

The resonant frequency is the oscillation of a system at its natural or unforced resonance. Resonance occurs when a system is able to store and easily transfer energy between different storage modes, such as Kinetic energy or Potential energy, as you would find with a simple pendulum. Most systems have one resonant frequency and multiple harmonic frequencies that get progressively lower in amplitude as they move away from the center.

Resonance describes the phenomenon of increased amplitude that occurs when the frequency of a periodically applied force (or a Fourier component of it) is equal or close to a natural frequency of the system on which it acts. When an oscillating force is applied at a resonant frequency of a dynamic system, the system will oscillate at a higher amplitude than when the same force is applied at other, non-resonant frequencies.

Frequencies at which the response amplitude is a relative maximum are also known as resonant frequencies or resonance frequencies of the system. Small periodic forces that are near a resonant frequency of the system can produce large amplitude oscillations in the system due to the storage of vibrational energy.

Resonance phenomena occur with all types of vibrations or waves, including resonance of quantum wave functions.

The angle of intersection of a wave with its axis

The energy quality is based on the shape of the wave. When we say the shape of the wave, we are referring to the proportion of the wavelength to the amplitude, irrelevant of the single measurement of frequency or wavelength. This quality can also be expressed by the angle that the wave crosses the central axis. In other words, the relationship of the amplitude to the wavelength is the main factor determining the energy quality effect of the wave on its environment, which could also be expressed as in the form of the angle of intersection of the wave with its axis.

This concept was mentioned in the book "Back to a future for mankind." by Ibrahim Karim (an architect who is a graduate of the Federal Institute of Technology, Zurich, Switzerland, and teaches at several universities). It was mentioned in a different context related to how Biogeometry assesses the effect of a wave on the subtle energy systems of its surroundings due to the quality of shape of that wave.

Biogeometry is using the energy principles of shape to qualitatively balance biological energy systems and harmonize their interactions with the environment. (At least this is how Ibrahim Karim explains it). It is worth mentioning that the concepts of Biogeometry are neither widely recognized nor widely accepted. However, here we just want to discuss the idea of the angle of intersection of a wave with its axis within the context of this book.

Remember, we just mentioned that we usually characterize oscillators by the sharpness of the resonance. The sharpness of the resonance is the ratio of the beats of the resonance and the frequency. And that the Q factor explains how fast energy decays in an oscillating system. It compares the frequency at which a system oscillates to the rate at which it dissipates its energy.

What if there is another factor that could be added to the beats of the resonance and the frequency? This would be the angle of intersection of a wave with its axis, which would determine the effect of the energy of the wave on its environment.

An example of that would be earth. We mentioned that the Q of the rotation of the earth is 10^7. It fulfills all of our requirements for resonance and oscillations. But we know as well that we can now observe the Earth's axis to be tilted 23.5 degrees from the plane of its orbit around the sun. But this tilt changes. During a cycle that averages about 40,000 years, the tilt of the axis varies between 22.1 and 24.5 degrees. Because this tilt changes, the seasons as we know them can become exaggerated. More tilt means more severe seasons—warmer summers and colder winters; less tilt means less severe seasons—cooler summers and milder winters. It's the cool summers that are thought to allow snow and ice to last from year to year in high latitudes, eventually building up into massive ice sheets.

This is a direct observation that the angle is affecting the environment and that the angle could actually be the main factor leading to this effect, not the Q factor.

The thinking goes here that resonance changes the amplitude of the wave, which leads to changing the angle of intersection of a wave with its axis leading to a different effect on the environment/field/system.

Here, we are discussing macroscopic, planetary objects.

Could this be applied to microscopic, quantum objects?

The answer is that we need to research it and test it, but in principle, this could be applied to microscopic, quantum objects. The angle could be affecting the system or the environment. Yet, to explore this a little bit more. Let me tell you the electron story.

Every electron could be considered as a little bar magnet with a north pole and a south pole. The magnetic moment for electrons is all the same length. Let's talk about the concept of preparing a state and detecting a state of that electron.

Let's say we want to prepare this electron in a configuration where its magnetic moment is pointing vertically upward. How do we do it?

We simply create a large magnetic field. For example, you bring a magnet, turn on some current, and make an electron magnet with a north pole and a south pole. The electron is supposed to be in the magnetic field, of course. We expect from our little bar magnet (the electron) that it will process around the magnetic field. But sooner or later, it will get rid of that processional energy as it will radiate electromagnetic radiation. It will radiate away its energy and come into the best energetic configuration pointing straight up, where the north pole of the electron is pointing towards the south pole of the magnet and the south pole of the electron is pointing towards the north pole of the electron the magnet. And if we do it with a big magnetic field, it will radiate the energy quickly and very quickly come to an equilibrium with the north pole of the electron pointing vertically upward.

We could prepare the electron into a different configuration by rotating the magnetic poles at an angle.

Now, we want to do a measurement to find out which direction the electron is pointing in. Now, we are sort of carrying a detection experiment. We can ask what the electron's angle is relative to the magnetic field?

To answer that question, we do the same thing as we did to prepare the electron. We stick it into a large magnetic field and wait for it to radiate some radiations. If this was a classical setup (following the laws of classical physics), one measure of the angle would be how much radiation the electron emitted. For example, if it was almost perfectly upward, to begin with, then we would expect that it only has a little bit of extra energy. Perfectly upward has the minimum energy. Rotate away a little bit; you give it a little bit of magnetic energy after letting it come to rest again; it does so by giving off that little bit of energy. If you were to go to a more extreme situation where you pointed the electron horizontally, then it would emit even more energy by the time it got vertical. Finally, if you point it straight down, it would emit the maximum amount of energy in the form of radiation. And then, you could at least measure the angle relative to the magnetic field by measuring how much radiation comes out. And the answer would be a nice continuous function of the angle of the electron. So you would measure the spin direction of the electron by letting it emit some radiation and counting up the amount of radiation.

It turns out that for a real electron, this is completely the wrong story!!

This is not the way it works!!

Something else happens, which illustrates the weirdness of the quantum world. It is not as we expected. It does not behave the way it should if it follows the laws of classical physics.

First of all, no matter which way you prepare the electron, then you turn on the magnetic field, and one of two things happens. Only two things could happen. One is that it emits no photon, no electromagnetic radiation. The other is that it emits one quantum, one photon of electromagnetic radiation of a very particular frequency that corresponds to the energy of jumping from down to up. It is almost as if the electron had only two possible configurations. There are only two states of an electron either it is pointing up, or it is pointing down. That is one of the puzzles of quantum mechanics.

Of course, there is a whole lot of mathematics and quantum mechanics dealing with that. Yet, here we want to examine the effect of the angle of intersection of the wave with its axis. Let`s consider the electron as a fluctuating wave. As mentioned above, it is expected that the angle of intersection of a wave with its axis is the main factor determining the effect of the wave on the environment/field/system. Following this line of thought, electrons and other fundamental particles should not be changing amplitudes and should not be changing the angle of intersection of a wave with its axis. In other words, since they are fundamental, the update of their wave function should not result in changing the angle of the intersection of a wave with its axis.

This is, again, of course, just an analogy aiming at examining the effect of the angle of intersection of a wave with its axis and whether this idea could be applied to sub-atomic objects with quantum mechanical behavior, and it will require further scientific research and validation.

Some of the proposed operations of the Balancing force (The B-force):

One of the potential operations of the balancing force is to bring fields or systems to a "Balanced state." The balanced state does not mean a zero value or a ground state. It is a state where the system is in balance to achieve its targeted function within a specific space and time.

$$B \ast f\hat{\Psi}(x,t) = |S_b\rangle$$

Where:

- B is the Balancing force.
- ✶ is the sign indicating that it is operating "in" something.
- $f\hat{\Psi}(x,t)$ is a field of waves the B-force is operating "in"
- $|S_b\rangle$ refers to the resultant of the operation, which is the balanced state.

The balanced state

The balanced state $|S_b\rangle$ would be the resultant of the balancing force operation on any operatee which could be a field of waves or just one single wave function.

The balanced state is usually not zero, and it should be the value or set of values that achieve the required balance for the operatee/system in which the balancing force is operating "in."

A very basic example for that would be our straight line analogy, where the balancing force is operating "in" the straight line construction of reality to maintain its balanced state $|S_b\rangle$ which is expected to be a wavelength or a length of 10^{500} in one dimension. A simple notation that could represent that:

$$B \ast \bar{x} = |S_b\rangle$$

Where:

- B is the Balancing force.
- ✶ is the sign indicating that it is operating "in" something.
- $\bar{x}$ is the operatee or the straight line.

- $|S_b\rangle$ refers to the resultant of the operation, which is the balanced state.

Later in this book, we propose how the higher dimension force (The HD-force) can operate on the straight line to reconstruct reality and the information coded on the straight line into two dimensions. However, the operation of the HD force could be independent of the operation of the B-force. They might not be affecting each other as the B-force is expected to be required to maintain the balance of the straight line, even if the HD force is operating on it.

The balancing force and the cosmological constant

Another example of the balancing operation of the B-force could be the cosmological constant.

From the perspective of quantum field theory, every point in space is represented by a quantum oscillator, one for each elementary particle type. Higher energy oscillations represent the presence of real particles. However, even the lowest possible energy oscillation, the one corresponding to the absence of particles referred to as a vacuum state, has some energy. To satisfy the Heisenberg uncertainty principle, the vacuum state of any field oscillation $E = \frac{1}{2}hv$, where v is the frequency of the oscillation.

The theoretical value of "zero-point energy" suggested by quantum field theory is 10^{112} ergs/cm³. This estimate was first made by John Wheeler and Richard Feynman, who noted that one teacup of space with this energy density would contain enough energy to boil all of the oceans on the planet.

The observed value of the vacuum energy density (the small value of the cosmological constant) is 10^{-8} ergs/cm³. This discrepancy of 120 orders of magnitude is referred to as the vacuum catastrophe or the cosmological constant problem.

If some fields can have extremely large positive zero-point energies, then perhaps others have extremely large negative zero-point energies that cancel them out. An extension of the standard model of particles called supersymmetry may partially allow this. It gives all particles a supersymmetric counterpart that may precisely cancel out the vacuum energy. Basic supersymmetry only allows us to cancel out photons down to the electroweak energy, which brings the predicted vacuum energy down to a mere 10^{47} ergs/cm^3. For a while, theorists assumed that something like this must be happening, meaning the vacuum energy was really zero.

In the late 90s, astronomers discovered that the expansion of the universe is accelerating in the way we would expect from non-zero vacuum energy. The observation of accelerating expansion allows us to measure the absolute density of vacuum energy.

Quantum field theory can give us extremely high vacuum energy or vacuum energy of exactly zero if we assume the symmetry of positive and negative zero points between different fields.

But a very small non-zero vacuum energy? That requires what is referred to as fine-tuning. Gigantic positive and gigantic negative zero-point energies cancel each other out down to a very tiny non-zero value.

The balancing force could be a proposed solution.

Earlier in this book, in the section "The physics behind the straight-line analogy," we considered a value of 6.22404×10^{-8} K to represent the Einstein temperature where all elements of the periodic table when shattered to their elementary particles form BECs, and there are no fluctuations. This could be referred to as the "Balanced state of the ISWs identity stem waves."

Since 10^{-8} is a value comparable to the vacuum energy density, we will consider that the vacuum energy density is the "balanced state of the vacuum."

The balancing force could be operating on the vacuum energy –in some form of operation that needs to be -and could be figured out to bring it to the "Balanced state" of 10^{-8}. This operation aims not to cancel gigantic positive and negative energies out but to bring them to a "Balanced state," which is estimated to require a value of 10^{120}.

The notation that was proposed to express this type of operation:

$$B \divideontimes f\hat{\Psi}(x,t) = |S_b\rangle$$

Where:

- B is the Balancing force.
- $\divideontimes$ is the sign indicating that it is operating "in" something.
- $f\hat{\Psi}(x,t)$ is a field of waves the B-force is operating "in"
- $|S_b\rangle$ refers to the resultant of the operation, which is the balanced state.

The non-zero principle of the balancing force

The balancing force cannot have a value of zero. The operations of the balancing force are targeting to bring energies or fields to a "Balanced state." Every system/field has its non-zero balanced state, which could be calculated.

The connection coefficients of the balancing force

Connection coefficients should allow the transformation of the balancing force to carry out the targeted operation, for example bringing a field down to its balanced state $|S_b\rangle$

A notation that could be used to express the introduction of connection coefficients to the balancing force operation:

$$B\,\Gamma^k_{ij} \divideontimes f\hat{\Psi}(x,t) = |S_b\rangle$$

Where Γ^k_{ij} is just a notation for the connection coefficient, which indicates that there are other components, factors, or forces that could affect the transformation or the operation.

The balancing force and the fine-tuning

Physics is full of constants; these are values or properties for which there is no theoretical basis. They are only determined by measuring them. Some of these constants are things like the charge of the electron, the gravitational constant, Planck's constant, and others.

If any of these numbers were different, our universe, as we know, would not be the same, and we probably would not be here if that was the case. Why do these numbers have the values that they do? Why these exact values make the universe suitable for us to be here? This is basically the core of the fine-tuning argument.

It goes something like this: The constants of nature are statistically improbable. These constants are on a razor's edge such that even a slight deviation would likely result in no life as we know it in the universe.

Some argue further and say that this fine-tuning is unlikely that it could not have occurred by pure chance. That there must be an intelligent designer that set up the constants to enable life. Others reject this argument for various reasons.

It is more or less like asking why we happen to live on a planet that happens to be at just the right temperature for liquid water to exist. That is a narrow edge; it is not as much of a knife-edge as the cosmological constant. The answer is that on planets where there can't be water, there can't be life. So, it's truly a very small fraction of planets at the right temperature for water to exist. Where do you live? In the only place we can live, where water exists.

The same kind of picture, the universe is vast, very diverse, with many different environments and possibilities. Among these possibilities in a few small pockets of the universe, the conditions are right for life, and that is where life exists. With planets, we know that there are many planets where a small number will have liquid water and the vast majority will not. With the universe, we rely on theoretical ideas to say that there are different bubble/pocket universes, each one of which has different laws. Possibilities are like blueprints. They are blueprints for different kinds of universes. This could be analogous to the possibilities of life, as explained by Leonard Susskind. DNA is the blueprint for life. DNA has a vast number of ways of being rearranged, so there are many possibilities for life, but that in itself does not say that there is an enormous number of living creatures around. It took something to make those blueprints into actual houses or whatever it happens to be.

Part of the story is cosmological, that the expansion of the universe, the inflation of the universe, the very rapid expansion that took place at very early times created a lot of quantum fluctuation, and that quantum fluctuation created patches of space with different properties, those patches are sometimes called pocket universes or sometimes called bubble universes. We live in one of them; that is the picture; there is mathematics that goes with it, which is sometimes referred to as multiverse.

Leonard Susskind used the term landscape to describe it, where he mentioned that the background comes from biology. The landscape of biological designs. And it means all the possible ways you can put DNA together; it is a tremendously large number of possibilities describing life. Similarly, all the various possible ways you could put the elements of physics together to make different kinds of pocket universes, just as the number of ways that you can arrange a DNA molecule is enormous. You have a DNA molecule with a billion base pairs. You can rearrange them in a massive number of ways, and that is why there are so many different possibilities for life. Similarly, the number of different blueprints is enormous.

An analogy that the science fiction writer Douglas Adams used is a sentient puddle. A puddle wakes up one morning after a night of rain and says, this is an interesting hole I find myself in; it fits me rather well. It fits me perfectly. It must have been made to have me in it. As the sun rises, the water evaporates, the puddle keeps thinking that the hole was made to have him in it until he disappears. But there could be lots of holes, with lots of sentient puddles with different dimensions of the hole.

Similarly, if we change the constants analogous to the dimensions of the holes, it would be an alternate universe. But life as we know it then may be different. And we would continue to say in that alternate existence. This universe must have been made to have us in it.

Our understanding of the universe is incomplete, and we know it is incomplete. We cannot yet explain dark matter, dark energy, the asymmetry of matter versus anti-matter, and a host of other things. It is quite possible that future, more complete theories could answer some of these questions and explain the theoretical basis for the constants being what they are.

Einstein said that nature shows us only the tail of the lion, but there is no doubt the lion belongs with it, and we see the lion only as a louse sitting upon him would. It is possible we simply have not yet seen the lion that awaits at the end of the tail.

Are constants of nature changing?

Paul Dirac, in 1937 speculated that physical constants such as the gravitational constant or the fine-structure constant might be subject to change over time in the proportion of the age of the universe. This has been a work in progress and an ongoing area of research.

In a paper published in the journal Science Advances, scientists from UNSW Sydney reported that four new measurements of light emitted from a quasar 13 billion light-years away reaffirm past studies that have measured tiny variations in the fine structure constant.

UNSW Science Professor John Webb says, "We found a hint that that number of the fine structure constant was different in certain regions of the universe. Not just as a function of time, but actually also in direction in the universe, which is quite odd if it's correct, but that's what we found."

The balancing force as a different perspective

The balancing force, along with the other imaginary complex forces proposed in this book, could give a plausible explanation from a different perspective for the idea of fine-tuning.

The thinking goes that the universe is in a "balanced state" rather than in a fine-tuned state. The balancing force drives this balanced state for a specific period and space to achieve particular conditions for different forms of life generation mechanisms.

There would be another operating force here, which would be the retrieving force that would be working on retrieving this kind of ecosystem to its original state. This is yet another important differentiation between the balancing force and the retrieving force.

This would be analogous to our sentient puddle mentioned above. The balancing force would be the reason for the dimensions and the equilibrium factors for our sentient puddle, and the retrieving force would be the evaporation of the water.

Both forces could be measured and potentially tested, and observed. The proposed mathematics and notations could be a tool to measure them and track their effects. Otherwise, a different mathematical framework could be used.

For example, the proposed symmetry-breaking mechanism for the identity stem wave of helium ISW_{He} could be a potential explanation for why the speed of light and massless particles has its value.

The higher dimension force (The HD force)

Some of the expected properties and characteristics of the higher dimension force (The HD force):

- **It is a phase transition and phase conversion accelerating force causing a paradigm shift:**
  This means that when the HD force operates on something or someone -either it operates on matter or organisms- it causes the acceleration of phase transition or even phase conversion. It changes the function of what it is operating on. This is not acceleration as the rate of change of velocity with respect to time. This is totally different and could have different mathematics describing it as well; this is basically accelerating the phase transition or conversion by introducing new information to the operatee (The thing the HD force is operating on) or, in other words, to the wave function of the operatee. This information will lead to a paradigm shift in the operatee and a change in its core function.

    Let's use a notation for that:

    $$\text{HD} \divideontimes f\widehat{\Psi}(x,t) = \frac{\partial E}{\partial t}$$

    - Where HD would be the higher dimension force.
    - $\divideontimes$ is the "In sign" indicating that the HD force is operating "In" something or someone.
    - $f\widehat{\Psi}(x,t)$ Refers to a complicated field of wave functions in a specific space and time. With a note here, that space and time $f\widehat{\Psi}(x,t)$ Are different from the space and time of the HD force. In other words, the HD force is independent of space and time parameters of the complicated field of wave functions $f\widehat{\Psi}(x,t)$ It is operating "in."
    - $\frac{\partial E}{\partial t}$ is just a notation to indicate the phase transition or conversation of the operatee (The thing the HD force is operating on).

This equation and notation are just indicative; it might contain mistakes, it might require some complex numbers, and Planck constant. It is just merely a basic description of what this paradigm-shifting operation might look like. A different notation could be used as well as the idea here is that there would be a change in the Energy field of the operatee it could be with respect to time.

For a very basic example of that, we will use the straight-line analogy introduced earlier in this book. That is a one dimension straight line. The HD force would be operating "in" our straight line. This could be either by introducing new information or reconstructing the information scrambled in the straight line differently. For example, the potential results of this operation could be a reconstruction of the one-dimensional information into two-dimensional information in the form of a tiny fluctuation on the straight line without affecting its geometry. It could be on the center of the straight line or at any other point on the straight line.

The source stimulating the tiny fluctuation of the straight line is expected to be outside the straight line. It is not limited by the straight line's dimensionality or space-time parameters. It could be a unified source for the three complex forces, which could be referred to as the "$HDBR_3$" for the Higher dimension force, the balancing force, and the retrieving force unified, which will be explained in further detail later in this book.

The HD operation could have a **"scaling & projection operation"** as well.

- **The scaling and projection operation of the HD force:**
  This could be one of the operations of the HD force, where it applies scaling mechanisms, scalar fields on an operatee along with a projection of the rescaled information into a different space which could be referred to as base space in some cases. It could just be a different space like space A, space B, etc.

Disclaimer: The following notations and equations are just a proposal to approach the idea of the operations of the HD-force, it borrows equations from the Higgs mechanism and symmetry breaking, yet other notations and approach could be more accurate, and the below method could contain glitches and mistakes that will require corrections.

Let`s continue with the example of the straight-line analogy and apply the scaling and projection operation of the HD force on it. Let`s consider the scalar field ϕ (x), (it is just a notation for the scalar field) which uses a solution to the classical equation ϕ_0 then we perturb small fluctuations about that solution ∂ϕ(x)

$$\phi(x) = \phi_0 + \delta\phi(x)$$

If we consider it in the following Lagrangian:

$$L = \frac{1}{2}\partial\mu\phi\partial^\mu\phi - v(\Phi)$$

For ϕ = Constant $\quad\quad\quad\quad\quad \frac{\partial v}{\partial \phi} = 0$

$$v(\phi) = -\frac{1}{2}\phi^2 + \frac{1}{4}\phi^4$$

The symmetry of the potential is that:

$$v(-\phi) = v(\phi)$$

We expect to get:

$$L(\phi(x)) = \frac{1}{2}\partial\mu(\phi_0 + \delta\phi)\delta^\mu(\phi_0 + \delta\Phi) \ldots\ldots$$

$$=$$

The derivative of the fluctuations – the potential v (ϕ) (Equation 2)

$$\frac{1}{2}\delta_\mu(\partial\phi)\delta^\mu(\delta\phi) + \frac{1}{2}(\phi_0 + \partial\phi)^2 - \frac{1}{4}(\phi_0 + \delta\phi)^4$$

For the solution ϕ = 0

$$L_0 = \frac{1}{2}\partial_\mu \partial\phi \partial^\mu \partial\phi + \frac{1}{2}\partial\phi^2 - \frac{1}{4}\delta\phi^4$$

This Lagrangian (the values of the equations of motion around ϕ) has the symmetry of ∂ϕ → -∂ϕ

The idea here (which could be derived or presented using a different mathematical framework and potentially more accurate approaches and notations) is that fluctuation(s) happen on the one dimension straight line without changing its geometry and applying a scalar field will be a tool to reconstruct the information and project it on a different space.

A simpler notation that could represent the idea of the HD force's scaling and projection operation:

$$\text{HD } \phi(x) \divideontimes \bar{x} = (\sigma : \text{SP}_m \longrightarrow \text{SP}_1) \longrightarrow \text{ISW}_n$$

This equation represents an example of a scaling and projection operation by the HD force, where:

- HD refers to the higher dimension force.
- $\phi(x)$ Refers to the scalar field or operator operating "in" something.
- ✻ is the "in sign" indicating that the HD force using a scalar field or operator is acting "in" the operatee.
- $\bar{x}$ here represents the straight line in our straight-line analogy, but more generally, it represents the operatee, so if it`s a two-dimensional operatee it could be $(\bar{x}, \bar{y})$. If it is a complicated field of waves (which could be the situation in many cases), it could be $f\widehat{\Psi}(x,t)$, so the proposed equation would be as follows:

$$\text{HD } \phi(x) \divideontimes f\widehat{\Psi}(x,t) = (\sigma : \text{SP}_m \longrightarrow \text{SP}_1) \longrightarrow \text{ISW}_n$$

For the other side of the equation:

- σ Represents a point on the main space or, more generally, the operatee.
- SP_m represents the main space.
- SP_1 represents the base space or space number one.
- ISW_n represents the ISW or the identity stem wave (Fluctuation) resulting from the operation; for example, it could be the ISW of iron, then we can refer to it as ISW_{Fe}

To solve our basic example of the straight-line analogy, we will use the basic form of the proposed equation

$$\text{HD } \phi(x) \ast \bar{x} = (\sigma : SP_m \longrightarrow SP_1) \longrightarrow ISW_n$$

For $\bar{x}$ we will consider a length of 10^{500}; it is just a number that we borrowed from the landscape of string theory (but it could be another number).

The expected result of the operation is scaling the 10^{500} into $5*10^{-5}$, which would be the identity stem wave ISW of iron, and again this is just an example. And the value of $5*10^{-5}$ would be projected from the main space to the base space or our space to form the first ISW of our two-dimensional super symmetrical crystal. So, one-dimensional straight line information was reconstructed through this operation into two-dimensional information and projected to a different space.

Ricci scalar and the scaling and projection operation of the HD force

The Ricci tensor tracks volume changes along geodesics. So, if we take an object and make all its points travel along geodesic curves, it shrinks in size. The Ricci tensor tells us about how volumes grow or shrink as they move along geodesics. There are two geometrical approaches to understanding the Ricci tensor sectional curvature and the volume element derivative.

The Ricci scalar is a number that is used to compare the volume of a ball in curved space to the volume of a ball of the same radius in flat space.

The Ricci tensor represents gravity in general relativity, as we can see here in the Einstein field equations:

$$\boxed{R_{\mu\nu}} - \frac{1}{2} g_{\mu\nu} + \Lambda g_{\mu\nu} = \frac{8\pi G}{C^4} T$$

The more space-time is curved, the more quickly bodies will get drawn together. In general relativity, gravitational attraction is the natural result of curved space-time. It does not require any forces, as we see in Newtonian gravity.

The Ricci scalar keeps track of how a size of a ball in curved space deviates from the standard size of a ball in flat space. A ball in flat 2D space would just be a circle. A circle will have the same perimeter in flat space and curved space. However, the circle in curved space can fit more area inside it because of the nature of curved space.

The Ricci scalar helps us quantify how much more space we can fit inside the circle in a curved space compared to the flat space circle.

So, through the power of curved space and the Ricci scalar, we can fit a lot more space inside a tiny boundary.

The Ricci tensor components are:

$$R_{k_j} = R^i_{kij}$$

When the Ricci tensor acts on the same vector twice, the result we get is the Ricci curvature.

The Ricci curvature can only tell us about volume changes. It cannot tell us whether or not a ball is changing shape. It will only tell us if the ball has changed in volume. The formula of the Ricci curvature

$$RiC(\vec{v},\vec{v}) = \sum_{i=1}^{D} k(\vec{e}_i,\vec{v})$$

Ricci flow equation corresponds to the structure getting smaller as you add more structure:

$$\partial g_{\mu\nu}(x) = -R_{\mu\nu}$$

The Ricci scalar/tensor could be a mathematical tool to track the operations of the HD force, compare and track the changes in volume caused by the HD force.

The Ricci flow could help us understand when you add more ISW to the ISW structure, it gets smaller until the ISW of Helium bangs and breaks the symmetry as proposed earlier in this book.

- **Conservation and the Higher dimension force (The HD force)**

    It is expected that the HD force does not violate any conservation laws. The way the HD force is expected to be operating on fields, systems, matter, and organisms would be adding new information and potentially some in the form of energy that could be referred to as the HD energy. These added information or energy are not governed by space and time limitations of the operatee. They do not necessarily cause changes in the volume or the shape of the operatee. They might do in some cases, but it is not necessarily an effect of this added information or energy. Then the laws of conservation would apply to the new information or energy introduced to the operatee.

- **The paradigm shift operations and effects caused by the HD force**

One of the main operations/effects of the HD force is expected to be a paradigm shift for the operatee. In order to explain the expected nature of these paradigm shifts, what should be considered as a paradigm shift caused by the HD force, and what should not, we will use the analogy of human beings and evolution.

For example, in this analogy, natural selection should not be considered as a paradigm shift caused by the HD force even if different species evolved from common ancestors and the new species are having new functions and have different abilities and skills. Still, this is driven by specific environmental and ecosystem conditions leading to evolution through natural selection. The paradigm shift that could be caused by the HD force, in this case, would be something that the conventional environmental and ecosystem conditions will not impact, for example, behavior like being ambitious, innovation, and seeking knowledge that is not required for your survival. In my previous book, "Quantizing Emotions- the Basic Mathematics of Psychology," some of these behaviors were introduced as what could be referred to as UCIB or (the Universal Collective Impact Behaviors), like seeking knowledge, sharing knowledge, and implementing knowledge in forms of innovation and invention.

The idea here is that biologically humans have followed the evolution process through natural selection. Yet, mentally or at the conscious level, they have faced a paradigm shift that could be through an operation by the HD force that we could potentially figure out its mathematical tools and mechanisms.

Another example of the paradigm shift caused by the HD force could be from our straight-line analogy. The force behind the first fluctuation/ripple on the straight line that reconstructed the information on the straight line from one dimension construction of reality to two-dimension reconstruction of reality causing a paradigm shift and reconstruction of reality.

- **Entanglement as a mechanism for the HD force.**

    Resonance is expected to be one of the mechanisms of the balancing force. Yet, for the HD force, entanglement would be an expected mechanism. And it is not spooky action at a distance. There is a force behind it that is not spooky or magical, and this force does not define distance or space, or time as we define it. But we can still figure out the mechanisms by which this force operates, and entanglement could be a very good candidate.

    Unitarity is the quantum equivalent that tells you that you can always reconstruct the past from the future. In the state of a quantum system, you can either run forward uniquely or run backward uniquely, and you will come to some unique previous state or future state. It is a kind of time reversibility.

    We lose information because we lose the ability to follow the details, not because the information gets lost. That is when the second laws come when you lose the ability to follow the information.

    The evolution of systems can be represented by matrices. Multiplying matrices by matrices, if we want to update a second time, apply the same matrix to the resultant. If you want to update a state of a system five units of time, multiply the matrices together five times, you do it in sequence.

    So, let's consider a system of a field of wave functions evolving as we would expect, and then a paradigm shift was introduced to it by the HD force, which could be through entanglement. We can track this paradigm shift by identifying which parameter(s) of the matrix's wave function(s) has changed. This is not an easy task, and it requires tracking these changes down to a very fundamental level.

Another analogy would be human beings as a highly sophisticated measuring apparatus for the universe with a very high Q-factor. The measurement process that we do leads to a collapse of the universe's wave function that we observe/measure. In other words, reality as we perceive it would be a projection from the brain on the fabric of space-time. An entanglement happens between us (the measuring apparatus) and the field of waves. The entanglement could be with a specific collapse of a wave function. This wave function is not necessarily bound with our space-time parameters. If we develop the mathematics of the HD force and the complex forces in general further, we could track and measure such entanglement. The two-worlds interpretation introduced in this book proposes another approach for this idea.

The mechanism of the HD-force's entanglement
The nature of the entanglement properties of the higher dimension force (The HD-force) is not necessarily always as quantum entanglement. Potentially for the mechanism of HD-force's entanglement:
- o The HD force's entanglement will not be limited to a specific number of states.
- o "Partial entanglement" could be allowed, which means that the two entangled systems could be partially entangled, in a sense that some of the properties of one of the systems or fields could be entangled with some of the properties of the other system or field. And not necessarily the entire system/particle. Entanglement entropy could be the mathematical tool to measure the degree of entanglement.
- o "Fading of entanglement" could be allowed and tracked as well. The term fading of entanglement would probably be more accurate than using "decay of entanglement."
- o The HD-force could choose to pass the information to an imaginary particle/field, which is not limited by our space-time parameters. Then this particle/field would be entangled with the particle/field limited by our space-time parameters.

The Notebook thought experiment

This would be just a thought experiment in an attempt to understand the suggested mechanism of entanglement of the Higher Dimension force (The HD-force), which is not necessarily quantum entanglement as suggested in the previous section. The analogy could be used to understand the mechanism of the "unified complex forces" $HDBR_3$, which is proposed in the next section of this book. Additionally, the analogy will help in understanding the two-worlds interpretation proposed in this book as well.

The analogy goes that we have a notebook of, let's say, 120 pages of holographic two dimensions films. These 120 pages would be representing our multi-verse, so all the possible universes are within this notebook. There might be other notebooks with different stories, but this is beyond the target of this analogy.

If Alice –which is standing outside the notebook- looks at the two-dimensional holographic page of our universe, she would not be able to tell what this thing is a hologram of unless she shines a light on it to see the three dimensions representation of reality of this holographic film.

Technically, Alice is outside the space-time of this holographic film, and she would be shining light from outside it. What if she wants to change the coding or the information on that holographic film? Alice from her location cannot change the code on the two-dimensional holographic film. Yet, according to the two worlds interpretation presented in this book, Alice has a second version of her in our universe's page on that holographic film. The second version of Alice can change the code using the mechanisms proposed in the straight-line analogy, project it on the total space, which will be projected back on the base space, which is our universe.

Some of the operations suggested for the HD-force like scaling and projection, in addition to entanglement, could be a good candidate for the tools she can use to shine a light on the two-dimensional holographic film as well as the idea of changing the code on that two-dimensional holographic film through her version in our universe. Now, this could as well start from a one-dimensional representation of reality, as suggested in the straight-line analogy.

Another question would inevitably come up, what if she wants to transfer codes, information, or even energy between the 120 pages?

Well, this could probably be done as well if we understand well the operations of the three imaginary complex forces the Higher Dimension force (The HD-force), The Balancing force (The B-force), and the retrieving force (The R-force).

The primary operations of these forces are:

- The Higher Dimension force (The HD-force): **Scaling, projection, entanglement, and paradigm-shifting.**
- The Balancing force (The B-force): **Resonance.**
- The retrieving force (The R-force): **Creation and annihilation.**

The imaginary forces can be unified and act together as well on an operatee, as explained in the next section of this book.

If we managed to reach a good understanding of which operation can be used to do which function, we would probably be able to transfer data, information, and potentially energy between the 120 pages of the notebook, in addition to shining the light and changing the code from outside the space-time of the notebook.

What is time in the notebook thought experiment?

From Alice's perspective, she just sees a two-dimensional holographic film in this notebook. Information is coded in two dimensions. The definition of time in these two dimensions is not clear for her or could even be irrelevant for her. If we refer to the two worlds interpretation presented in this book, Alice would have another version of her in the universe coded on this holographic film. This information is reconstructed in three dimensions (or four dimensions); when Alice shines a light on it and starts moving the light along the holographic film, then things start to happen, and time starts getting a relativistic meaning.

Alice is not controlling her version of our universe; she merely sheds light on the holographic film to represent it in three dimensions of space and give time meaning. Neither does Alice has the ability to change the code. However, her version in our world can change the code, as explained in the straight-line analogy.

From Alice's frame of reference, everything is happening/happened now, Including the Big Defreeze (if proven correct). The Big Bang, inflation, formation of galaxies, evolution, and the end of the universe. If she shined a light at the holographic film's correct part, she would see one part of the movie. However, Alice is not interested in that as her main job is shining a light for her version of our universe to get represented in three dimensions. She might not have the ability as well to shine a light except on the section of the holographic film where her second version's information is coded.

Bob happens to be shining a light on the holographic film for his version of our universe to get represented in three dimensions simultaneously when Alice is shining a light. Time in their frame of reference is different from how their versions in our universe sense time. Basically, their versions in our universe would perceive time as a function of the flow of events which gets represented in three dimensions as a result of shinning the light from their other versions.

An almost unavoidable question comes up, what happens to their versions in our universe when they die?

Without getting too philosophical, if we are following the same logic of this thought experiment, this would mean that their versions in the second world would stop shining light on the holographic film to reconstruct their information in three dimensions. But that does not mean that the information of their versions in our universe got erased from the holographic film.

Charlie comes at a later stage and shines a light on his version of our universe to reconstruct it in three dimensions. But he cannot shine a light on Alice`s and Bob`s versions because he can only shine the light on the section of the holographic film where his second version`s information is coded.

Could the information of Alice & Bob be reconstructed? And if they do get reconstructed, would they have quantum mechanical constituents? Would they have the ability to measure/observe the other scenarios so that these scenarios will actually happen or at least some of them? These are some open questions to think about.

The caveman analogy

Let`s travel back in time to meet a caveman & give him a glimpse of the future with a nice movie. We will leave the TV with the caveman for a while and observe how he deals with it.

What would be his expected reaction?

He might get scared and run away and never come back. Then he might start scratching his head and try to understand what is going on. He will get closer to the TV, touch it, and hit it. He will search for the source of the people in the movie, he will try to talk to them with no response from their side, and they will continue doing whatever they are doing.

The caveman might notice few cables connected to the TV. He will pull one cable, the sound will go away with more gestures of curiosity and surprise, and our caveman might plug the cable again, or pull another cable, now the people in the movie disappeared. He will plug the cable again; the people will come back. The caveman will not understand what is going on, and there is no hope that he will get it during his lifetime unless he gets a revelation or something. Even with that, he might still not understand the complete technicalities of how this is operating.

The best guess he could make is that the source of the people and the sounds on the TV are the cables he is pulling out and plugging in, which would be considered as an intelligent observation compared to the people of his time, and they may assign him as the king of the tribe. He might take it further and say that these cables create the people and the sounds on the screen, to be then tagged as the greatest philosopher of his time thanks to his genius observations and conclusions. Ego might hit him harder, and he can claim that he can manage the death and resurrection of the people in the movie by pulling out and plugging in the cables, and he might claim that he is the God of this tribe.

We know that the TV is merely a tool for receiving and transmitting waves of information; this information has been recorded in a different time and location. The source of this information is the writer, director, and actors in that movie from a different place and a different time.

We decided that we want to stop our friend, the caveman, from brainwashing his tribe and try to explain to him what is going on. The guy is quite intelligent, and surprisingly he is almost philosophically outsmarting us by a shocking reply:

Why do you guys not be receivers and transmitters too? And the wires in your brains are like the cables in the TV. If you pulled out one of them, it would affect some of your functions, but they are merely cables helping you receive and transmit the information. Yet one TV can transmit high-quality sound and pictures, giving you a great experience. Other TV can ruin the sound and the picture and, in a sense, pollute them.

Now we are the ones scratching our heads with gestures of surprise and confusion. We decided to ask him if that is the case, what would be the source of information?

He gets closer to us, looks us in the eye, and then looks at the sky and says confidently: The universe; there could be more than one, it could be a multiverse.

We decided to pack our TV and tools, get on board with our time traveling machine and go back to our people in the future; we have some interesting stuff to report.

Chapter 05: The two-worlds interpretation

The two worlds interpretation
Let's start with a brief about the many-worlds interpretation of quantum mechanics, the problems with it. Then we will present the idea of the two-worlds interpretation based on the thought experiments and analogies introduced in this book.

Here is a summary of what the many-worlds interpretation says:

One of the strangest features of the quantum description of reality is the idea of superposition. We cannot describe the most fundamental building blocks of our universe with defined singular properties. Instead, they seem to behave as "probability clouds" of all properties they might have. Mathematically, this is encapsulated in the wave function of a quantum particle or system of particles.

The double-slit experiment is an illustration of why we need to describe the quantum world this way. Briefly, a stream of photons or electrons or even molecules travels from some point to a detector screen via a pair of slits. These particles arrive at the screen distributed like the interference pattern you would expect from a simple wave. Quantum mechanics successfully predicts this result by describing each particle's journey as a superposition of all possible trajectories. In other words, the particle simultaneously takes all possible paths, which means it passes through both slits. It tries out all histories between launch and landing. And those many "maybe" histories somehow interact with each other to determine the most likely final destination when a measurement is made. In a sense, different possible superposed histories appear to converge on one final outcome. But what causes that convergence?

In the original Copenhagen interpretation of quantum mechanics, the act of measurement was thought to collapse possibility space into a single reality at least with respect to the measured property. It collapses the wave function. That collapse signifies the transition between the quantum and the classical realms.

One of the founders of quantum mechanics, Erwin Schrödinger, criticized this. And he proposed his famous Schrödinger's cat thought experiment to highlight the absurdity. Briefly, it goes like this. A cat is in a box with a flask of poison. A machine containing a radioactive element is set to shatter the flask in the event that the radioactive element decays. If that happens, the cat dies. That radioactive decay is a purely quantum process, so it exists in a superposition of states until it is observed. It has both decayed and not decayed.

But does not that mean that the entire macroscopic system attached to that quantum event is also in a superposition? If so, then the cat should be simultaneously alive and dead until we open the box. But why can't the cat collapse its own wave function? And from its point of view, is the physicist outside a "quantum blur" until the box is opened? And what about the entire rest of the universe that physicists or cats are not currently observing?

Many adherents to Copenhagen's interpretation propose a resolution to the paradox of Schrödinger's cat. It is that quantum superposition does not extend to macroscopic scales. It disappears when different quantum scale histories diverge. This is called decoherence when the wave function describing quantum systems overlaps sufficiently. In other words, they are coherent. It is possible to get interference in the double-slit experiment and correlated entanglement measurements. But when these systems interact with their environment, coherence is lost, and parallel histories fall out of alignment. They can no longer interact with each other. By Copenhagen's interpretation, we might say that the universe chooses the final outcome of all those histories. It does not exactly choose a single history. Instead, it chooses an end result, say particle location on a screen or cat alive or deadness, based on those histories. Suppose a larger number of those histories lead to a given result. In that case, it's more likely that the universe will select that outcome. Copenhagen's interpretation says that this selection happens in a fundamentally random way. It is what we would call a non-deterministic interpretation because there is no underlying predictability behind the selection.

However, there is another way to interpret the transition between the quantum and classical worlds. What if the wave function never collapses? If we can imagine a cat in a superposition of states, alive and dead, why stop at the cat? What if the family of possible states extends beyond the radioactive decay, beyond the cat, and includes the observer and indeed the entire universe too? If we open the box and find that the cat is alive, it is because we are part of an entire quantum timeline in which the radioactive decay and subsequent poisoning never happened. But there is an equally valid timeline in which it did and another version of us experiencing that. This sounds outrageous, but it is a serious interpretation of the mathematics of quantum mechanics. It was proposed by Hugh Everett in his 1957 Ph.D. thesis entitled "The Theory of the Universal Wave Function." It has come to be known as "The many-worlds interpretation."

To outline the idea, let's talk about what this means in the context of the double-slit experiment. The Copenhagen interpretation tells us that the superposition of particle trajectories (of histories) merges into the single timeline of the observer's reality. Many worlds say this merging never happens. Those alternative histories continue, and we find ourselves in just one of those timelines.

The many-worlds interpretation invites the idea that reality splits into different branches every time quantum states diverge into different possibilities, for example, at every particle interaction, everywhere in the universe. This would lead to an unthinkably large number of alternate timelines or "worlds" that contain all possible realizations of this universe since the Big Bang.

The superposition of states of many worlds can be thought of as over laid histories, slices of a universal wave function that diverge from each other as the universe evolves, but none ever vanish.

Many worlds may imply that every possible version of you exists out there. You are just the one who happens to be experiencing this "branch of reality." Every other possible live path, including those branching in different directions from every decision you ever made, may be just as real.

Some of the problems with the many world interpretation

In quantum mechanics, every system is described by a wave function from which one calculates the probability of obtaining a specific measurement outcome. From this wave function, you can calculate, for example, that a particle that enters a beam splitter has a 50% chance of going to the left and a 50% chance of going to the right. But once you have measured the particle, you know with 100% probability where it is.

This means that you have to update your probability and, with it, the wave function. This update is called the wave function collapse.

This wave function collapse is not optional. It is an observational requirement. We never observe a particle that 50% here and 50% there. Speaking of 50% probability makes sense as long as we are talking about a prediction.

Now, this wave function collapse is a problem for the following reason. We have an equation that tells us what the wave function does as long as we do not measure it. It is called the Schrödinger equation, which is a linear equation. This means that if you have two solutions to this equation and you add them with arbitrary pre-factors, then this sum will also be a solution to the Schrödinger equation. Such a sum is also called the superposition.

The problem is now that the wave function collapse is not linear, and therefore it cannot be described by the Schrödinger equation. Suppose you have a wave function for a particle that goes right with 100% probability. Then you will measure it right with 100% probability. Likewise, if you have a particle that just goes left, you will measure it left with 100% probability. But, here is the thing. If you take a superposition of these two states, you will not get a superposition of probabilities. You will get 100% either on one side or the other. Therefore, the measurement process is not only an additional assumption that quantum mechanics need to reproduce what we observe. It is actually incompatible with the Schrödinger equation.

Now, the most obvious way to deal with, that is to say, well, the measurement process is something complicated that we do not yet understand and that the wave function collapse is a "placeholder" that we use until we figure out something better. Yet, most sign up for the Copenhagen interpretation, which in a sense says that you are not supposed to ask what happens during measurement. In this interpretation, one could argue that quantum mechanics is merely mathematical machinery that makes predictions, and that is it.

The problem of the Copenhagen interpretation – and similar interpretations- is that that you require you to give up the idea that what a macroscopic object like a detector does should be derivable of its microscopic constituents. In a sense, the Copenhagen interpretation implies that what the detector does cannot be derived from the behavior of its microscopic constituents.

The many world interpretation suggests that every time you make a measurement, the universe splits into several parallel worlds, one for each possible measurement outcome. In the many-worlds interpretation, if you set up a detector for a measurement, then the detector will also split into several universes. Therefore, if you ask what will the detector measure? Then the answer is that the detector will measure anything that is possible with probability one. This, of course, is not what we observe. We observe only one measurement outcome. The adherents of the many world interpretation explain this that you are not supposed to calculate the probability for each branch of the detector, because when we say detector, we do not mean all detector branches together. You should only evaluate the probability relative to the detector in one specific branch at a time.

Quantum mechanics tells us that matter is made of elementary constituents that are often referred to as particles, but they are actually described by wave functions. A wave function is a mathematical object that is neither a particle nor a wave, but it can have properties of both. The curious thing about the wave function is that it does not itself correspond to something which we can observe; instead, it is only a tool that helps us to calculate what we do observe. To make such a calculation, the quantum theory uses the following postulates and properties:

1. As long as you do not measure the wave function, it changes according to the Schrödinger equation.

    - The Schrödinger equation is different for different particles, but its most important properties are independent of the particle.
    - One of the important properties of the Schrödinger equation that it guarantees that the probabilities computed from the wave function will always add up to one.
    - Another important property of the Schrödinger equation is that the change in time which you get from the equation is reversible.
    - The Schrödinger equation is linear. This means if you have two solutions to this equation, then any sum of the two solutions with arbitrary pre-factors will also be a solution.

2. The second postulate of quantum mechanics tells you how to calculate from a wave function the probability of getting a specific measurement outcome; this is called the Born rule, named after Max Born, who came up with it.

    - The Born rule says that the probability of a measurement is the absolute square of their part of the wave function, which describes a certain measurement outcome.

- To do this calculation, you also need to know how to describe what you are observing, for example, the momentum of a particle.

3. **The measurement postulate** is sometimes called the update or the collapse of the wave function. This postulate says that after you have made a measurement, the probability of what you have measured changes to one. This is a necessary requirement to describe what we observe.

    - If you do not update the wave function after measurement, the wave function does not describe what we observe. We do not ever observe a particle that is 50% measured.

The problem with the quantum measurement

The problem with the quantum measurement is that now the update of the wave function is incompatible with the Schrödinger equation.

The Schrödinger equation is linear, which means if you have two different states of a system, both of which are allowed according to the Schrödinger equation, then the sum of the two states is also an allowed solution.

The best-known example of this is Schrödinger's cat, a state that is a sum of both dead and alive; such a sum is what physicists call a superposition. We do, however, only observe cats that are either dead or alive. This why we need the measurement postulate; without it, quantum mechanics will not be compatible with observation.

The measurement problem is not solved by decoherence.

Decoherence is a process that happens if a quantum superposition interacts with its environment. The environment may be air or even in a vacuum; you still have the cosmic microwave background radiation. There is always some environment. This interaction with the environment eventually destroys the ability of quantum states to display typical quantum behavior, like the ability of particles to create interference patterns. The larger the object. The more quickly its quantum behavior gets destroyed.

Decoherence tells you that if you average over the states of the environment –because you do not know what they do- then you no longer have a quantum superposition. Instead, you have a distribution of probabilities. This is what physicists call a mixed state. This does not solve the measurement problem because, after the measurement, you still have to update the probability of what you have observed to 100%. Decoherence does not tell you to do that.

One of the issues with the measurement postulate is that the behavior of a large thing like a detector should follow from the behavior of the small things that it is made up of. But this is not the case. So that is one of the issues; the measurement postulate is incompatible with reductionism. It makes it necessary that the formulation of quantum mechanics explicitly refers to macroscopic objects like detectors, when really what these large things are doing should follow from the theory.

Some people seem to think that you can solve this problem by a sort of reinterpreting the wave function as encoding the knowledge that an observer has about the state of the system. This what is called a Copenhagen or Neo-Copenhagen interpretation.

Now, if the wave function merely describes the knowledge an observer has, then you may say, of course, it needs to be updated if the observer makes a measurement, which is very reasonable. But this also should refer to macroscopic objects like observers and their knowledge. And suppose you want to use such concepts and postulates in your theory. In that case, you are implicitly assuming that the behavior of observers or detectors is incompatible with the behavior of the particles that make up the observers of detectors. This requires that you explain when and how this distinction is to be made.

The many-worlds interpretation does not solve the measurement problem briefly because it has to use a postulate about what a detector does, as explained earlier in this section.

The two-worlds interpretation
To help understand the idea of the two worlds interpretation and how it could be a good candidate for a solution for the measurement problem, we will go back to the notebook thought experiment introduced earlier in this book.

The analogy goes that we have a notebook of, let`s say, 120 pages of holographic two dimensions films. These 120 pages would be representing our multi-verses, so all the possible universes are within this notebook, including our universe.

If you look at the two-dimensional holographic page of our universe, you would not be able to tell what this thing is a hologram of unless you shine a light on it to see the three-dimensional representation of reality of this holographic film. Technically, you are outside the space-time of this holographic film, and probably you are not governed by the laws of this holographic film, including the laws of quantum mechanics. You would be shining light from outside it.

The idea is that we have two worlds. The first one is in our universe, on that two-dimensional holographic film having quantum mechanical behaviors. The other one is totally outside "the notebook" (our universe or multiverse) and its space-time parameters. It does not follow the quantum mechanical behavior.

The version of you in our universe is an observer with constituents following the quantum mechanical behavior, which requires the second version of you in the other world with constituents that do not have quantum mechanical behavior.

Even if the idea of the multiverse is eventually proven correct, according to the proposed two worlds interpretation, the multiverse will not necessarily imply that there are many versions of you doing all possible scenarios. The two worlds interpretation suggests that there is only one version of you acting as an observer with constituents that do not have quantum mechanical behavior. The second version is acting as an observer of the first version. But the second version does not require an observer because its constituents do not have quantum mechanical behavior.

Many questions, yeah?

The great thing about science and thought experiments is that when you figure out one conclusion or follow a line of thought that you think will lead you to explain everything. It almost instantly opens doors for further research, keeping us busy and thinking about the quest to get closer to the truth. And as I always say: **"Science is the crawling of humanity towards the truth."**

Yet, when we discuss an idea like that, we need something testable, reproducible, and, if possible observable. We know that the many world interpretation is not yet observable. Yet, the mathematics of quantum mechanics suggests the idea of the many-worlds interpretation.

The idea of the two-worlds interpretation could potentially be a candidate to solve the measurement problem, which could potentially be mathematically verified. Additionally, the idea of the imaginary complex forces, their properties, and operations introduced in this book could provide a good basis for developing more sophisticated mathematics for the two-worlds interpretation. Furthermore, toward the end of this book, we will propose some ideas for testing and observing the idea.

The two-worlds interpretation also implies that when an act of measurement happens by the first version, which has quantum mechanical behavior, a simultaneous act of measurement is required by the second version in the other world that does not have quantum mechanical behavior.

So back to some of the questions that may come up about the two-worlds interpretation:

- Where is that second world/version if it is outside the multi-verse space-time?
- What laws does it follow if it does not have quantum mechanical behaviors?
- How does it interact with our world? Does the second version impact the first version? And how?
- Could the first version impact the second version (its observer version)? And how?
- Why is an HD-entanglement operation required with our observer version?
- What would be the source of knowledge/information of your other version in the second world?
- Why do we need all this loop of encoding information between the two worlds in the first place?
- What happens to all the other possible scenarios?
- What would the second version be doing?

Of course, not all the answers are available yet, and there could be other questions too. Yet, at least we can try to approach the solutions for some of these questions.

Where is that second world/version if it is outside the multi-verse space-time?

The potential answer could be based on the straight-line analogy introduced in this book, where we have three different representations or reconstructions of reality in one dimension (straight line), in two dimensions (holographic film), and in three dimensions (or four space-time dimensions). The information is coded and scrambled into the two-dimension holographic film. Additionally, it is coded and scrambled into one dimension straight line.

The thinking goes that the other world is even outside the space-time parameter of the straight line. Space and time are defined differently in that world; they could potentially have a totally different meaning, they could be meaningless as well in an attempt to visualize it, which brings us to the next question.

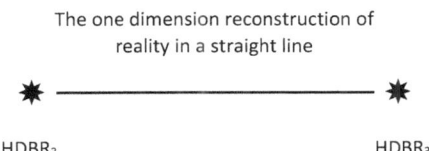

What laws does it follow if it does not have quantum mechanical behaviors?

The potential answer to this question could be based on the imaginary complex forces, their properties, and operations introduced in this book. Which might not give the full answer, but at least something to begin with.

The three imaginary complex forces are the Higher Dimension force (The HD-force), The Balancing force (The B-force), and the retrieving force (The R-force).

The main operations of these forces are:

- The Higher Dimension force (The HD-force): **Scaling, projection, entanglement, and paradigm-shifting.**
- The Balancing force (The B-force): **Resonance.**
- The retrieving force (The R-force): **Creation and annihilation.**

It is important to emphasize that the entanglement operation of the HD force does not necessarily behave as a quantum entanglement. This brings us to the next question.

How does it interact with our world? Does the second version impact the first version? And how?

It is essential to highlight that your observer version in the other world is a version of you that does not deterministically tell you what to do or sort of control and order you what to do or what not to do, nor does it control your fate or decisions. But without getting philosophical, it would be merely an observer with constituents that do not follow quantum mechanical behavior, yet it might have some sort of impact on you.

The potential answer to this question could be based as well on the properties and operations of the imaginary complex forces like scaling and projection, entanglement, resonance, creation, and annihilation. It is essential to highlight that this will mostly not fully follow the quantum mechanical behavior. Yet, the mathematics of quantum mechanics could provide a basis for understanding these sorts of interactions. The introduction of the "in sign" to the Schrödinger equation as explained in this book and my previous book "Quantizing emotions – The basic mathematics of psychology" could also be a good basis for understanding this interaction. Briefly:

$$i\hbar \frac{\partial}{\partial t} \psi(x,t) = \hat{E} \; ✳ \; \hat{\psi}(x,t)$$

Here the equation is prescribing that the energy operator is acting "in" a wave function, which is the operatee in this case, and this wave function does not have definite energy; it could be a field of waves. This equation also prescribes that the energy operator and the wave operatee affect each other and can swap roles that are basically what the "in sign ✳" indicates.

The three imaginary complex forces could be unified and help in these kinds of interactions between the two worlds which will be explained in further detail in the next section of this book.

The mechanism of the entanglement between these two worlds does not necessarily follow a quantum entanglement behavior. Please check the mechanism of the entanglement of the HD force introduced earlier in this book. Briefly:

The nature of the entanglement properties of the higher dimension force (The HD-force) is not necessarily always as quantum entanglement. Potentially for the mechanism of HD-force`s entanglement:

- The HD force`s entanglement will not be limited to a specific number of states.
- "Partial entanglement" could be allowed, which means that the two entangled systems could be partially entangled, in a sense that some of the properties of one of the systems or fields could be entangled with some of the properties of the other system or field. And not necessarily the entire system/particle.
- "Fading of entanglement" could be allowed and tracked as well. The term fading of entanglement would probably be more accurate than using "decay of entanglement."
- The HD-force could choose to pass the information to an imaginary particle/field, which is not limited by our space-time parameters. Then this particle/field would be entangled with the particle/field defined by our space-time parameters.

Could the first version impact the second version (its observer version)? And how?

This would be a bit tougher question to answer. Yet, the potential answer could be in the topological manifolds, and manifold bundles mentioned earlier in this book, specifically the C-line bundle. Briefly:

If we consider a map σ that starts in the base manifold M and maps into the total space ($\sigma : M \longrightarrow E$), so it is not in the projection direction, it is the other direction. Such a map is called a section of the bundle if you can apply the map σ to go from M up to the total space, and then you apply the projection π afterward, and if this leads you back to the identity on the base space id_M then you have a section. ($\pi \circ \sigma = id_M$).

Yet, here if we consider the straight line as the total space and our world/universe is the base space. We still have an issue: our observer version is outside the space-time parameter of the straight line/the total space, so how would the first version impact it? And the potential answer to this would be in the HD-force entanglement mechanism, which is not necessarily quantum entanglement.

The thinking goes that when an action σ is mapped from our base space, in other words, our world/universe to the total space (the straight line) and coded in that straight line, an HD-entanglement operation could happen with our observer version.

Why is an HD-entanglement operation required with our observer version?

Some people seem to think that you can solve the measurement problem in quantum mechanics by a sort of reinterpreting the wave function as encoding the knowledge that an observer has about the state of the system. This what is called a Copenhagen or Neo-Copenhagen interpretation.

Now, if the wave function merely describes the knowledge an observer has, then you may say, of course, it needs to be updated if the observer makes a measurement, which is very reasonable. But this also should refer to macroscopic objects like observers and their knowledge. And suppose you want to use such concepts and postulates in your theory. In that case, you are implicitly assuming that the behavior of observers or detectors is incompatible with the behavior of the particles that make up the observers of detectors. This requires that you explain when and how this distinction is to be made.

The two-worlds interpretation could potentially be a good candidate to solve this problem as well. The question would be the wave function would be describing the knowledge of which observer? Is it our world's observer version of you who has constituents that behave quantum mechanically and observe objects that behave quantum mechanically or your observer from the other world who has constituents that do not behave quantum mechanically?

The thinking goes that if the version of you in our world is observing some other object in our world, then the constituents of you and the object you are observing have quantum mechanical behavior. It is then expected that the wave function would be describing your knowledge and the wave function needs to be updated. The assumption here is that the behavior of observers or detectors is actually compatible with the behavior of the particles that make it up. They both have quantum mechanical behavior.

But let's consider the two-worlds interpretation. An act of measurement/observation is required to be done on you by your other version in the other world with constituents that do not have quantum mechanical behavior. Yet, at least it should be having knowledge about the state of the system that the wave function is encoding/updating. Then an inevitable question comes up?

What would be the source of knowledge/information of your other version in the second world?

The short answer would probably be the first version of you in our world/universe is the source of this knowledge/information.

This quite counter-intuitive. We proposed earlier that the entanglement operation is not necessarily quantum entanglement. We highlighted some of the potential properties of the entanglement process and how it is potentially differentiated from quantum entanglement, which could be the operation between the two versions of you. In that case, we should be open to the idea that the mathematics of this operation might not be limited to the mathematics of a wave function. Rather it should represent the entanglement operation and any other potential imaginary complex forces operation that could be introduced to the process. This implies that the mechanism of encoding the information is not necessarily described by the mathematics of the wave function.

The basic proposed mathematics in this book for the imaginary complex forces could be a good candidate to start. Otherwise, the mathematical logic could be built differently to describe the idea.

The thinking goes that there is a process of encoding the information between the two worlds and the two versions of you with a mechanism that is yet to be better understood and analyzed. There is also a probability that this process could be non-linear. Here is a proposal of how the process might go:

- The knowledge/information would be projected from the version of you in our world/universe (the base space) to the total space represented in the straight line. An entanglement process then happens between this point of information (let's call it a map σ for now) and the second version of you.

- This information (map σ) is sent because you are doing an act of measurement in our universe where you and the object you are measuring have quantum mechanical behavior. In this case, a simultaneous act of measurement on you is required (since you need an observer) by the second version of you in the other world. Yet, the second version of you does not require an observer because it does not have quantum mechanical behavior.

- The knowledge/information is then encoded through an operation or a set of operations –that is yet to be fully understood- but it could contain the scaling and projection, entanglement, creation, and annihilation operations by the imaginary complex forces proposed in this book. The system or the wave function is then updated with the new information.

- If we consider that a wave function is involved, it will then be encoding the knowledge/information of your observer (the second version of you). This knowledge/information was initially projected from the first version of you on the base space to the total space. There is a chance as well that different mathematics could be used for different phases of the operation. In other words, as the operation evolve, it might initially require to use the mathematics of the imaginary complex forces. Then as it evolves further, at a point in this operation, it can start satisfying the wave function mathematics.

The question is simply why?

Why do we need all this loop of encoding information between the two worlds in the first place?

The answer could potentially be that simply because *we need a mechanism for the first version of you in our world/universe to be able to change the code which is reconstructed in different dimensions without any sort of interference or control either from the second version of you in the other world or from any other forces*. And that mechanism requires such type of operations. This could be a deep philosophical discussion and could have deep philosophical meanings. Yet, what we are concerned with for the purpose of this book is to explore the potential mathematics of these operations and see how physics could describe it in a potentially testable and reproducible manner.

What happens to all the other possible scenarios?

This is a fundamental question, especially when comparing the many-worlds interpretation with the two-worlds interpretation.

The short answer is that they get erased; they never happen.

And the reason could potentially be decoherence.

Erasing the other scenarios here does not mean destroying the wave function or collapse of the wave function. It simply means updating the wave function in a sort of formatting all the other possibilities to either have an updated wave function with no information or with different information.

Let`s explain this proposal in further detail; we will start with decoherence.

Decoherence is a process that happens if a quantum superposition interacts with its environment. The environment may simply be air, or even in a vacuum, you still have the radiation of the cosmic microwave background. There is always some environment. This interaction with the environment eventually destroys the ability of quantum states to display typical quantum behavior, like the ability of particles to create interference patterns. The larger the object, the more quickly its quantum behavior gets destroyed.

Decoherence tells you that if you average over the states of the environment —because you do not know exactly what they do- then you no longer have a quantum superposition. Instead, you have a distribution of probabilities. This is what physicists call a mixed state. This does not solve the measurement problem because, after the measurement, you still have to update the probability of what you have observed to 100%. Decoherence does not tell you to do that.

The argument is that in the context of many-worlds interpretation, there is no wave function collapse at all; decoherence explains how we lose the ability to see these alternate histories. The multiple branches of the wave function as it interacts on macroscopic scales. We only see what remains visible to us, stranded as we are on a single branch of the universal wave function that itself contains so much more than our little decohered slice of space-time. Yet, as mentioned above, the interaction of the quantum superposition with its environment eventually destroys the ability of quantum states to display typical quantum behavior.

The entanglement of the detector/observer and the system it measures is a case of entanglement. That is what a measurement is; an entanglement between the detector/observer and the system it is measuring.

So we have two entanglements here. The first one is the quantum entanglement between the observer (your version in our world) and the system you are observing/measuring. Both the observer and the system have quantum mechanical behavior. The second entanglement –which is not necessarily quantum entanglement, as explained earlier- is between your version in our world and the second version in the other world that does not have quantum mechanical behavior.

The first entanglement, which is quantum entanglement between your version in our world and the system, happens when you do the measurement/observation. The wave function is updated, and we have 100% probability.

Yet, another operation(s) is expected to happen here; it is related to what happens to the "probability cloud" of all the other possibilities if we want to get away from the concept of the collapse of the wave function. One of the potential ways to prevent the idea that everything that could happen will happen in a parallel universe and to get away from the collapse of the wave function is introducing other operation(s) that could be carried out on this "probability cloud" and all other possibilities to sort of erasing them.

As explained earlier, erasing the other scenarios here does not mean destroying the wave function or collapse of the wave function. It simply means updating the wave function in a sort of formatting all the other possibilities to either have an updated wave function with no information or with different information.

This process of erasing the other scenarios or updating the wave function so that other scenarios lose or transform the information within them could potentially happen through the operations of the imaginary complex forces introduced in this book, including annihilation and creation, resonance, entanglement, scaling, and projection.

These operations are likely to be carried out by your second version in the second world who does not always have quantum mechanical behavior. The reason is that if you carry out these operations and your constituents have quantum mechanical behavior, then every time you do a measurement, everything that could happen will happen because of quantum mechanics. But suppose your second version in the second world, whose constituents do not always have quantum mechanical behavior, do these operations. In that case, quantum entanglement will not happen, and the ability of quantum states to display typical quantum behavior will be destroyed. The wave function will be updated or, in other words, sort of formatted by erasing the information of all the other scenarios.

So if we considered the two-worlds interpretation, Elvis Presley would not be still alive in other universes singing in one of them and studying physics in another one. Yet, only a second version of Elvis Presley could still be alive in the second world. The question is, what would he be doing? Is he still singing?

What would the second version be doing?

What we mainly need to know about the second version of ourselves – at least for the purpose of this book- is what the role of this second version is. However, in terms of how she/he is spending their time, this is an area to explore on future occasions.

As explained earlier, we proposed that the second version is not within the space-time of our multiverse; we proposed as well that the constituents of the second version do not always have quantum mechanical behavior. It seems that the main role of the second version is to observe the first version, not for surveillance purposes but to enable the first version to carry out her/his own observation and measurement without interference or control, as explained earlier. It is a sort of mandatory requirement to have a non-quantum mechanical observer to close a potential quantum mechanical endless loop of observers. This observer is proposed to be the second version of you in the two-worlds interpretation.

In a sense, the second version of you in the second world would be like a puppeteer holding the strings for you but not controlling you. You would be giving him the directions on how to move the strings. (By you, I mean the first version of you in our world). The reason that the second version of you is required to hold the strings and vibrate them as per your instructions is that the puppet's world can only work with strings.

Another important role of the second version of you in the two-worlds interpretation is to update the wave function of the "probability cloud", all the other probabilities/scenarios in a formatting manner. In other words, your second version will be erasing all the other scenarios by formatting the wave function and updating it.

Chapter 06:

The Unified complex forces HDBR$_3$

According to modern particle theory represented by the standard model of particle physics, all matter is composed of six quarks and six leptons and their 12 anti-particle pairs. But that is not all; matter is subject to 4 fundamental forces that cause this matter to behave in specific ways and result in essentially every action you see all around you and everywhere in the universe.

These four forces are the strong force that binds the atoms' nuclei by holding protons and neutrons together in their center. The weak force is responsible for some kinds of radioactivity. Electromagnetism is responsible for electricity and light. And finally, gravity binds us to the earth and keeps the planets in their orbits around the sun.

Yet, one of the exciting aspects is that scientists believe that all the forces come from one underlying force or principle at a fundamental level. Not only that, but according to current understanding, all 24 of the different fundamental particles and the four forces are one and the same at some deep level. This is the ultimate symmetry of the universe.

Perhaps the best way to understand this and how these forces emerged is to visualize what happened at the very beginning of time at the big bang when everything was one. Time began not at zero but at the smallest measurement of time that our models of quantum mechanics can represent, and that is Planck time 10^{-43} seconds. This is called the Planck epoch; all the forces and particles are one and the same at a point smaller than the size of a proton. Gravity is thought to have separated from everything else shortly after this time. So it was the first force to separate out from the other three forces. The temperatures at this stage of the universe were 10^{31} Celsius, and the energies were in the range of 10^{19} GeV.

The next era, called the grand unified epoch, lasts from the first Planck second 10^{-43} seconds to about 10^{-35} seconds; up to this period, the remaining three forces, the strong force, the weak force, and electromagnetism, were all unified. But shortly after this period, from about 10^{-34} seconds to about 10^{-32} seconds, the strong force separated from the other two electromagnetism and the weak force which were united as one force called the electroweak force; at this stage, temperatures were at around 10^{26} Celsius, and the energy was reduced to 10^{15} GeV. Somehow this separation of the strong force is thought to have resulted in or powered cosmic inflation.

At 10^{-12} seconds, called the quark epoch, the electroweak force split into the weak force and electromagnetism. So at this point, all the four forces became distinct. The temperature of the universe cooled to 10^{15} Celsius, and energies are about 100 GeV. The Higgs field exists at this stage as well. How do we know? Because 100 GeV was needed to create the Higgs boson, and this can be done in the LHC.

The HDBR₃ and the four fundamental forces

The following will be just analogous and will be built on the big defreeze & the straight line analogies presented earlier in this book.

So let`s consider our imaginary three "Complex forces," the Higher dimension force (The HD-force), the Balancing force (The B-force), and the Retrieving force (The R-force) unified as one force we will name it for short the HDBR₃

Let`s apply the HDBR₃ to our straight-line analogy.

The analogy goes that we had the straight line as a one-dimensional representation of reality. A tiny fluctuation happened on that straight line –without changing its geometry- that was projected on another space forming the ISW crystal. When all matter is shattered to their elementary particles, form BECs, which start fluctuating, forming what is referred to as the identity stem waves until the winding frequency of Helium`s ISW causes symmetry breaking, leading to the big bang and inflation.

Earlier in this book, we mentioned that one scenario could be the higher dimension force was the cause of this tiny fluctuation/ripple on the straight line. Yet, it is likely as well for the purpose of this analogy that HDBR₃ can operate together in some cases and independently in other cases. The HDBR₃ would be outside the boundaries of the straight line, probably at the beginning and the end of the straight line.

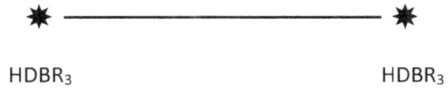

HDBR₃   HDBR₃

As mentioned earlier, entanglement could be one of the mechanisms of the Higher dimension force (The HD-force). Which, in our analogy here, could be the mechanism of how the HD-force could stimulate a fluctuation/ripple or more on the straight line. Either in the center of the straight line or at any other position. The next step would be carrying out a scaling and projection operation, which is one of the operations of the HD force presented earlier.

The scaling and the projection operation of the HD force would lead to more than identity stem wave ISW crystal, which would "give birth" to several universes; our universe is one of them.

If we consider the straight line as the main space and our universe as the base space, scaling and projection can happen from the main space to the base space and vice versa. It could happen in the other direction as well, from the base space to the main space.

The spectrum of energies of the $HDBR_3$
The following will be just an analogous approach to the idea.

Let`s consider a key of 12 energy qualities that covers the full energy spectrum. Let`s call this key HDBR12. We will consider that the wave has a horizontal component and a vertical component. The HDBR12 would represent the horizontal component of the wave. It would have opposite direction energies, representing the vertical component of the wave; we can use the notation the -HDBR12* to represent it.

The balancing force would have 7/12; these seven qualities have vibrational properties and obey the laws of resonance. As presented earlier, resonance is proposed to be the main mechanism of the balancing force.

The Higher dimension force would have 3/12. It will have three components in it. One component is purely related to the HD force. Another component would be its manifestation/entanglement/effect on the balancing force. The third component would be its manifestation/entanglement/effect on the retrieving force. Connection coefficients could be used to track the effect of the HD force on the balancing and retrieving forces.

The retrieving force would have 2/12. And that it is comparable to the spin properties. The retrieving force analogously has two states/spins. And this indicates that the retrieving force is the main force dealing with the quantum mechanical nature of our universe.

Let`s have a deeper dive into the 7/12 of the balancing force obeying the laws of resonance, and they are expected to have symmetrical properties as well. Let`s use the following notation for them:

- BE1A: Balancing energy 1A
- BE2A: Balancing energy 2A
- BE3A: Balancing energy 3A
- BEC: Balancing energy central
- BE1B: Balancing energy 1B
- BE2B: Balancing energy 2B
- BE2C: Balancing energy 2C

Balancing energy would be a result of the balancing force.

To help analyze the effects and the properties of this spectrum of energies of the balancing force, we will go back to the original equation proposed in the book "Quantizing emotions – the Basic Mathematics of Psychology."

$$\hat{E} \ast \hat{s} = f\hat{\Psi}(x,t)$$

This notation simply tells us that energy in shape gives a certain function; the in sign ✳ indicates the interchangeability of roles between the operator and the operatee.

We will consider that the shapes of objects are an expression of all the energy principles within them.

The above-proposed equation application for the balancing force was proposed to be:

$$B \ast f\hat{\Psi}(x,t) = |S_b\rangle$$

Where:

- B is the Balancing force.
- ✳ is the sign indicating that it is operating "in" something.
- $f\hat{\Psi}(x,t)$ is a field of waves the B-force is operating "in"
- $|S_b\rangle$ refers to the resultant of the operation, which is the balanced state.

A notation that could be used to express the introduction of connection coefficients to the balancing force operation:

$$B\,\Gamma^k_{ij} \ast f\hat{\Psi}(x,t) = |S_b\rangle$$

Where Γ^k_{ij} is just a notation for the connection coefficient which indicates that there are other components, factors, or forces that could affect the transformation or the operation.

So the proposed notation would be:

$$B\,\Gamma^k_{ij} \ast \hat{s}\, f\hat{\Psi}(x,t) = \sum_{BE}^{n} \ast |S_b\rangle\, f\widehat{\Psi}(x,t)$$

Where:

- B is the Balancing force.
- Γ^k_{ij} is the connection coefficient
- ✳ is the sign indicating that it is operating "in" something.

- ○ $\hat{S}$ is the shape of the operatee, which affects the operation (the balancing force could lead to a change in the shape and geometry of the operatee).
- ○ $f\widehat{\Psi}(x,t)$ is a field of waves or wave function the B-force is operating "in"
- ○ $\sum_{BE}^{n}$ ✹ is the sum of the balancing energy resulting from the left side of the operation and acting "in" the resulting field of wave/wave function.
- ○ |S_b> refers to the balanced state.
- ○ $f\widehat{\Psi}(x,t)$ is the resulting field of waves/wave function.

Even though this proposed equation has lots of notations, yet it is simply saying that when the balancing force acts on a field of waves, it gives a balancing energy acting on the resulting, updated system. We need to factor in the shape/geometry of the operatee and other factors that could be considered with the balancing force in the transformation process.

The in sign ✹ on both sides of the equation indicates the freedom to swap roles, $\hat{S}$ In some cases could be moved to the second part of the equation, as the balancing force could lead to a change in shape/geometry. And it could be affected as well by the shape/geometry of the field of waves it is operating in.

For the connection coefficient Γ_{ij}^{k}, the k, i & j are just indications of the components/factors that could be affecting the transformation process. For example, k could be the HD force, i & j could be the retrieving force.

The reason we have the field of waves/wave function $f\widehat{\Psi}(x,t)$ on one side of the equation and then the updated version of the field of waves/wave function $f\widehat{\Psi}(x,t)$ on the other side of the equation is that the updated field of waves/wave function $f\widehat{\Psi}(x,t)$ could again affect the state of the original field of waves/wave function $f\widehat{\Psi}(x,t)$ by the scaling and projection operation. The notation used in the equation describes that, and the in sign ✹ indicates the interchangeability of roles and effects between both sides of the equation. For further details about the in sign ✹, the logic behind it, how it originated, and the operations it indicates, please refer back to the book "Quantizing Emotions – The Basic Mathematics of Psychology," which was published on 17th of February 2021.

Again, a totally different mathematical framework & notations could be used to describe these concepts. Complex numbers and Planck`s constant might be required in some of these operations.

The HDBR$_3$ operations

Let`s consider that our imaginary "Complex forces," the HDBR$_3$ the Higher dimension force (The HD-force), the Balancing force (The B-force), and the retrieving force (The R-force), are unified. They can operate independently, and they can operate collectively.

For a collective operation of the unified HDBR$_3$ forces

$$\text{HDBR3 } \Gamma_{ij}^k \divideontimes \hat{s} \, f\Psi(x,t) = \Sigma_{HDBR3}^n \divideontimes |S_b\rangle \, f\widehat{\Psi}(x,t)$$

Where:

- HDBR3 is the unified complex forces
- Γ_{ij}^k is the connection coefficient
- ✹ is the sign indicating that it is operating "in" something.
- $\hat{s}$ is the shape of the operatee, which affects the operation (the balancing force could lead to a change in the shape and geometry of the operatee).
- $f\Psi(x,t)$ is a field of waves or wave function the B-force is operating "in"
- Σ_{HDBR3}^n ✹ is the sum of the spectrum of energies of the HDBR$_3$ on the left side of the operation and acting "in" the resulting field of wave/wave function.
- $|S_b\rangle$ refers to the balanced state.

- $f\widehat{\Psi}(x,t)$ is the resulting field of waves/wave function.

This equation is explaining the collective operation of the HDBR₃ on a field waves/wave function. There could be connection coefficients Γ_{ij}^{k} which allows for transformation operation. The equation indicates as well that the shape/geometry of the operatee could affect and could be affected by the operation.

Some of the independent operations of the HDBR₃ explained in details earlier:

- **For the retrieving force (The R-force):** creation and annihilation of space and time.

$$Ra_n^+ f\widehat{\Psi}(x,t) = \widehat{E} \divideontimes \hat{s}$$

- **For the Balancing force (The B-force):** Using the resonance mechanism to reach the balanced state of the system.

$$B\,\Gamma_{ij}^{k} \divideontimes f\widehat{\Psi}(x,t) = |S_b\rangle$$

- **For the Higher dimension force (The HD-force):** Scaling and projection operation using the entanglement mechanism.

$$HD\,\phi(x) \divideontimes f\widehat{\Psi}(x,t) = (\sigma : SP_m \longrightarrow SP_1) \longrightarrow ISW_n$$

Given that the HDBR₃ can operate independently or collectively, the HDBR₃ can be introduced to any of the independent equations. These proposed equations are just notations to explain the idea of the operations. Any of the components of these equations can be introduced to the other equation. The interchangeability between the independent and collective operations of the HDBR₃ is inevitable.

The fundamental differences between the proposed balancing and retrieving forces

There are fundamental differences between the balancing force and the retrieving force, some of the proposed fundamental differences:

- The retrieving force's main operation is to retrieve systems to their original state.
- The balancing force's main operation is to bring a field of waves to its balanced state. The original state could be the original state. Yet, it is expected that in most cases, the balanced state is different from the original state.
- In case the original state is the balanced state, both forces could operate collectively.
- One of the main mechanisms of the retrieving force is through creation and annihilation, which includes the creation and annihilation of space and time.
- The main mechanism of the balancing force is through resonance.
- Considering the HDBR12 key of the spectrum of energies, the retrieving force would have 2/12 with spin properties and operating with quantum mechanical rules.
- Considering the HDBR12 key of the spectrum of energies, the balancing force would have 7/12 operating with the laws of resonance in some cases independent of space and time boundaries or under different space and time parameters.

Back to some of the potential collective operation of the HDBR$_3$

$$\text{HDBR3 } \phi(x) \, \Gamma_{ij}^k \, a_n^+ \divideontimes \hat{s} \, f\widehat{\Psi}(x,t) = \Sigma_{HDBR3}^n \divideontimes |S_b\rangle \, f\widehat{\Psi}(x,t)(\sigma : SP_m \longrightarrow SP_n)$$

Where:
- HDBR3 is the unified complex forces
- $\phi(x)$ is the scaling and projection operator
- Γ_{ij}^k is the connection coefficient
- a_n^+ is the creation operator, it could be an annihilation operator a_n^-
- $\divideontimes$ is the sign indicating that it is operating "in" something.
- $\hat{s}$ is the shape of the operatee, which affects the operation (the balancing force could lead to a change in the shape and geometry of the operatee).
- $f\widehat{\Psi}(x,t)$ is a field of waves or wave function the B-force is operating "in"
- $\Sigma_{HDBR3}^n \divideontimes$ is the sum of the spectrum of energies of the HDBR$_3$ on the left side of the operation and acting "in" the resulting field of wave/wave function.
- $|S_b\rangle$ refers to the balanced state.
- $f\widehat{\Psi}(x,t)$ is the resulting field of waves/wave function
- σ is a point or section on the base space
- SP_m is the main space
- SP_n is the nth space

What this equation is telling us is that the unified HDBR$_3$ forces, the Higher dimension force (The HD-force), the Balancing force (The B-force), and the retrieving force (The R-force) are acting collectively on a field of waves/wave function using their mechanisms and operators including scaling and projection, connection, resonance and transformation, creation and annihilation. The shape of the operatee could affect and get affected by the operation. The resultant is the sum of the spectrum of energies resulting from the operation acting "in" an updated field; this field could be in a balanced state. The operation would lead to scaling and projection of a point or a section from the main space (it could be another space, not necessarily the main space) on the base space (or any other space). The equation is highly interchangeable, and some of the operators could be isolated depending on the operation. Planck's constant might be required to be introduced in some cases.

Chapter 07:
Testability of the ideas introduced in this book

The important question, can we test the new ideas and concepts introduced in this book?

Let's check how testable are some of the main concepts and ideas introduced in this book, and let's start with the Big Defreeze.

Testability of the Big Defreeze

The Big Defreeze briefly suggests that at the beginning of the universe, all matter started with a single collective quantum wave which could be referred to as the Identity Stem Wave of the Universe (ISWU). In other words, the universe was in a "frozen state" at almost absolute zero temperature and potentially shrunk in a very tiny collective wave, the "Identity Stem Wave of the Universe" ISWU. This stem wave started gaining energy leading to fluctuations behaving in the manner of a string forming what could be referred to as the "Identity stem waves of matter," leading to a super symmetrical crystal.

Of course, colliding all the matter that we know of or at least one of each matter to shatter them to their elementary particles. And then cooling the shattered particles to degrees close to absolute zero to see if they will form Bose-Einstein Condensates would be quite challenging, to say the least. And if we managed to do so, we will potentially be creating a new Big Bang. However, this sort of what is going to happen with the Big Rip scenario, where all the matter that we know off will be ripped. Then the next step, as proposed here, would be the heat death of the universe. So basically, the Big Freeze scenario would be modified a bit; it will basically be a scenario following the Big Rip.

Where after all matter is shattered/ripped, the universe will head towards the Big Freeze, and the shattered particles will form Bose-Einstein Condensates leading to a collective quantum wave. Which is proposed to be similar to the collective quantum wave -referred to in this book as the ISWU or the identity stem wave of the universe- at the beginning of the universe before inflation.

For Bosons and ideal gases, we know how to do experiments on them to form Bose-Einstein Condensates. The challenge would be for the matter, which is neither Bosons nor ideal gases; we will potentially need to do a fermion to boson transformation as suggested in the fermions problem earlier in this book.

The cold atom lab could be a very good candidate to carry out such experiments. Again it's quite challenging, but in principle, if we managed to break down matter to its elementary particles in the cold atom lab and then cool them down to near absolute zero and see what happens, we can start undertaking the behavior of the ISWs. We can also introduce two different types of matter, then three different types of matter, and so on.

Something to consider here, though, should we see how many elementary particles these matters are composed of and try to sort of isolate this specific number and configuration of elementary particles and introduce it in the cold atom lab? Or do we need to do the actual colliding and shattering of particles?

This is something that needs to be tested; the first method could yield good results. However, it is expected that the results will not be the same as if you actually collide and shatter the actual matter that you want to test. And this will not be because of some spooky action of knowing the history of the tested matter. It would be simply because of the idea of the evolution of information/wave function of this matter which is expected to have an impact on the result even if we use the exact same number and configuration of the elementary particles of this matter. Again, there are experimental challenges in that, but it is something worth considering as well.

Another point that is worth considering is the interaction with the environment. Even if you carry out the experiment in a vacuum, you still have the radiation of the cosmic microwave background. There is always some environment. Yet, let's consider the proposed identity stem wave super symmetrical crystal at the beginning of the universe. It was only interacting with its constituents, which are expected to be the newly born identity stem waves only. That was a very "pure" environment, and that was the only environment without any other interactions or cosmic microwave radiation. This means that potentially we can create identity stem waves and study their behavior. However, in a sense, we cannot create the same environment as the one of the proposed identity stem wave crystal at the beginning of the universe before symmetry breaking and inflation.

The Big Defreeze in the Cosmic Microwave Background

The idea of the Big Defreeze of the Universe suggests that the first identity stem wave that started fluctuating is that of iron ISW_{Fe} with a frequency of approximately $5*10^{-5}$. Then other ISWs started fluctuating until the ISW of helium ISW_{He} broke the symmetry of the ISW crystal with a value comparable to the speed of light.

In the 1990s, more precise measurements found that the Cosmic Microwave Background (CMB) is not perfectly smooth, but it has ripples in it. Inflation theorized that these ripples are the result of quantum fluctuations that were amplified when the universe expanded.

The temperature fluctuations $\Delta T/T = 10^{-5}$. These tiny fluctuations can be mapped. When you look at the structure of these fluctuations, you can Fourier decompose them and then calculate the power spectrum of these fluctuations; you find that you can get evidence from how these tiny fluctuations form, that for instance, most of the gravitating matter in the universe is actually invisible, it is dark matter we only see it via its rotational effects among others on how these fluctuations were formed. It is also the stuff that keeps the galaxies together; if we do not have this dark matter, they will just fly apart, and they will rotate too fast. And we also have evidence of dark energy which basically behaves like the cosmological constant, which has been now measured. It speeds up the expansion of the universe. It accelerates it.

If you take these tiny fluctuations, they cause variations in temperature and also in the density of matter and, therefore, variations to the gravitational potential. You put it in a computer run it forward; you find that these tiny fluctuations generate all the large-scale structures of the galaxy clusters and superclusters that we see today in our universe. You can show that wherever you find deviations from the mean temperature, you find clusters or voids. You can map this.

When you Fourier decompose it and compute the power spectrum of the two-point function of these fluctuations, you see that there are distinct structures in there. There are waves, sound waves basically, that propagated through the hot plasma before it cooled enough that hydrogen could form.

When you analyze the properties of these fluctuations, you see that – besides all the local dynamics like sound waves, the way how the sound waves interact with the plasma, the way how the plasma interacts with the photons, etc. – overall, you find a very remarkable structure. **You find that the waves that caused these tiny fluctuations in the temperature were generated with having the same power on all wavelengths. So they have what is referred to as the scale-invariant power spectrum. And they always come with an initial coherent phase condition as if they are coming from a broadband laser.**

This is considered evidence that there was a very early phase - much earlier than these 400,000 years after the Big Bang on the CMB- of the expansion of the universe that was much faster, exponentially fast, and this called inflation.

The Big Defreeze introduced in this book predicts that the source of this coherent, scale-invariant fluctuation is the identity stem wave of iron ISW_{Fe}. So, when the symmetry was broken by the identity stem wave of Helium ISW_{He} – as predicted in this book as well-. The ISW of iron maintained its power on all wavelengths as it did during the ISW crystal phase in maintaining its power on all the other ISWs leading to the interactions in the ISW crystal.

It is like, the ISW of iron acts as the source of the laser beam in the CMB, and it maintained its power and properties, including coherence and scale invariance. What was the source of the initial fluctuation in the ISW of iron? Please refer to the straight-line analogy and the proposed imaginary complex forces introduced in this book.

This would be evidence for the Big Defreeze if we managed to shatter iron to its elementary particles and carry out a fermion to boson transformation, then cool them down to near absolute zero to notice if a BEC would form. Then if we found that the Tc to form this BEC is comparable to $5*10^{-5}$. Then we have experimental evidence for the Big Defreeze. Despite the experimental challenges of actually shattering iron to its elementary particles, it is expected that it would yield more accurate results to do it, rather than considering how many elementary particles iron has and cool them down. And as explained earlier, this will not be because of some spooky action of knowing the history of the tested matter. It would be simply because of the idea of the evolution of information/wave function of this matter which is expected to have an impact on the result even if we use the exact same number and configuration of the elementary particles of this matter.

As it is clear, the idea of the Big Freeze is totally compatible with inflation and the Big Bang, with a tiny difference which is that the very first phase of the universe, which was tiny and condense was not hot; it was cold. And then, it got heated to a very high temperature through the symmetry-breaking mechanism of Helium`s ISW.

The era following the symmetry breaking by the ISW of helium leading to the bank is predicted to have the following operations:

- The transformation from two dimensions to three dimensions could be tracked using Riemann curvature and Ricci scalar, as explained earlier in this book.
- The identity stem waves of all matter ISWs were sort of freed and accelerating with speed comparable to the speed of light.
- Interactions happened between waves. And waves started gaining mass through the Higgs field.

- The ISWs started gaining more energy/heat; the acceleration is leading to an exponential increase in energy/heat and exponential expansion.
- The proposed imaginary mother forces will start acting on these waves with their operations, including creation and annihilation, resonance, scaling, and projection. With a note that these forces existed before the ISW crystal was formed.

How can we observe and track this in the Cosmic Microwave Background?

The answer would potentially be:

- The tiny fluctuations in the Cosmic Microwave Background.
- Operations of the imaginary complex forces.
- Evolution of information.

The Big Defreeze predicts that the tiny fluctuations in the Cosmic Microwave Background could potentially be the only fluctuation that maintained its full identity, coherence, behavior, properties, including scale-invariance during the first 370,000 years of the Universe and till date. These fluctuations correspond to the identity stem wave of iron ISW_{Fe} in the Big Defreeze.

That does not mean that the other ISWs lost their properties; basically, they are proposed to be "identity" stem waves which means they preserve –at least partially- their identities and properties. What it means is that the ISW of iron or the tiny fluctuations is predicted to be the only ISW that has fully maintained all its properties and power, and that is what we can currently observe on the tiny fluctuations in terms of having the same power on all wavelengths.

This prediction could potentially give us a great tool in tracking these fluctuations and their behavior and interactions during the 370,000 years. As it would be the only wave function that behaved exactly the same.

Laser cooling, optical lattice, and optical traps for testing the ISWs

The techniques used for the experiments to detect the BECs, interference of two BECs, and transferring them, especially the ones developed by Wolfgang Ketterle and his colleagues, could be very useful in testing the idea of the identity stem waves and their interactions.

Since the ISWs are basically a type of BEC, we expect to detect them and study their behavior and interactions mainly using the same techniques used for experiments on the BECs. We can do that now for bosons and ideal gas, which would potentially be sufficient to detect how far these predictions are correct. The main challenge would be for non-bosonic matter shattered into their elementary particles and transforming fermions to a bosonic quantum behavior like the Bose-Einstein condensation. We have proposed potential solutions for that in the "Fermions problem" section in this book. Another challenge would be the actual shattering of matter to maintain its history of information.

Laser cooling

The principles of laser cooling are pretty simple: you shine laser light on atoms, the atoms scatter light, and if you play some tricks, the light that is scattered/emitted has a shorter wave length and is more energetic than the absorbed light. The scattering of light removes energy from the system, and the system cools down.

Magnetic traps

There are two types of magnetic traps. Some have a pointy potential, a linear V-shape potential. And others have a round potential at the bottom; those traps are more tightly confining.

Interference pattern experiments

Let`s go back to the interference experiment by Ketterle and his colleagues, where you build two condensates, so you basically have lots of atoms in the ground states in each of the two traps, and then you turn off the traps, let them go. Then you can measure the distribution of atoms as a function of position. They basically saw fringes in the atomic density because the two condensates were interfering with one another.

The same experiments could be applied to the predicted ISWs; we can do that easily with bosons and ideal gas.

The optical trap could be used as a transport mechanism for the identity stem waves as well.

Testability of the imaginary complex forces

You can think of the proposed imaginary forces as imaginary numbers. They are important to make the mathematics work, you do not test them, and they are not real numbers.

Or can we experimentally test complex numbers?

In a paper entitled "Quantum physics needs complex numbers," published in January 2021, the paper proposes a way to experimentally test if the complex number exists, which is quite interesting. The details of the paper and the link for it are in the references.

In quantum mechanics, we work with complex-valued wave functions, and the equation that tells us what the wave function does in the Schrödinger equation, which has an i in it, which shows that the wave function has to be complex-valued. However, you can take the wave function and this equation apart into a real and an imaginary part. Indeed, one often does that when solving the equation numerically. If we calculate a prediction for a measurement outcome in quantum mechanics, then that measurement outcome will also always be a real number. So it looks like you can get rid of the complex numbers in quantum mechanics by splitting the equation into a real and an imaginary part, and that will not make a difference to the result of the calculation.

What they asked in the paper mentioned above is what happens with the wave function if you have a system composed of several parts?

In the simplest case, that would be several particles; in standard quantum mechanics, each of these particles has a wave function that is complex values. From these, we construct a wave function for all particles together, which is also complex-valued. What this function looks like depends on which particle is entangled with which. If two particles are entangled, this means their properties are correlated, and we know experimentally that this entanglement correlation is stronger than what you can do without quantum theory.

The question they look at in the paper mentioned above is whether there are ways to entangle particles in the standard, complex quantum mechanics that you cannot build from particles described entirely by real-valued functions. Previous calculations show that this could be done if the particles came from a single source. But in the paper, they look at particles from two independent sources and claim that there are cases in which you cannot reproduce with real numbers only. They also propose a way to measure this specific entanglement experimentally.

For that purpose, they considered scenarios in which there is more than one source of entangled states distributed to several parties who perform joint measurements. Such scenarios correspond to the future quantum internet, connecting many quantum computers and guaranteeing quantum confidentiality over continental distances. While Bell experiments have earned considerable fame, they just represent a simple instance of a quantum network consisting of one source and many observers. It is now well understood that networks with richer geometries, in particular including more than one source, offer new perspectives for Bell-type demonstrations.

So, according to the paper, if we propose a network corresponding to a standard **entanglement-swapping scenario** consisting of two independent sources and three observers, Alice, Bob, and Charlie. The two sources prepare two maximally entangled states of two qubits, the first one $\bar{\sigma}_{AB_1}$ distributed to Alice and Bob, and the second one $\bar{\sigma}_{B_2C}$, to Bob and Charlie. Bob performs a standard Bell-state measurement on the network scenario separating real and complex quantum physics. In standard quantum physics, two independent sources distribute the two-qubit states $\bar{\sigma}_{AB_1}$ and $\bar{\sigma}_{B_2C}$ to, respectively, Alice and Bob, and Bob and Charlie. At Bob`s location, a Bell measurement of four outputs is implemented. Alice and Charlie apply the complex measurements leading to the maximal violation of the CHSH$_3$ inequality: three and six measurements with two possible outputs, labeled by ±1.

The observed correlations cannot be reproduced or even well approximated when all the states and measurements in the network are constrained to be real operators of arbitrary dimensions. The impossibility still holds if the two preparations are correlated through shared randomness.

This measurement has the effect of swapping the entanglement from Alice and Bob and Bob and Charlie to Alice and Charlie: namely, for each Bob's four possible outcomes, Alice and Charlie share a two-qubit entangled state. Note that the actual state depends on Bob's outcome, but not its degree of entanglement.

Intuitively, the previous real simulation strategy could not work in this scenario. Using just four outputs, the measurement by Bob should swap not only the entanglement in the states but also the correlations in the extra flag qubits so that Alice and Charlie perform the original measurements or their complex conjugates in a correlated way. Clearly, this only shows that a given strategy for real simulation does not work. However, as mentioned and proved in the paper, it can be proven that no real simulation is indeed possible; that is, no real states and measurements can reproduce the statistics obtained in this entanglement swapping scenario.

The idea of designing experiments on scenarios in which there is more than one source of entangled states distributed to several parties who perform joint measurements is a good candidate for testing the imaginary complex forces proposed in this book. Where you have more than one source of particles/waves, and you have entanglement-swapping scenarios.

So, if we consider a quick analogy that we humans are detectors/measurement apparatuses moving in the universe and getting entangled with what we observe on a daily basis in continuous entanglement-swapping scenarios. This line of thought basically would be a more accurate description of nature.

Yet, we will then need the imaginary complex forces to carry out the proposed operations in this book. This book proposes the basic mathematical machinery to describe these proposed forces and the basic idea of potential designs of experiments to observe them.

Muons for testing the imaginary complex forces

In June of 2020, a collaboration of more than 170 scientists completed an extraordinarily complex set of calculations and arrived at this value: 2.00233183620(86) for the muon's g-factor.

On April 2021, the U.S. Department of Energy's Fermi National Accelerator Laboratory (Fermilab) announced the long-waited first results for the Muon g-2 experiment. The theoretical values for the muon g-factor are 2.00233183620(86), anomalous magnetic moment: 0.00116591810(43). The new experimental world-average results announced by the Muon g-2 collaboration are: g-factor: 2.00233184122(82), anomalous magnetic moment: 0.00116592061(41).

Let's consider that a string holds the muon in a magnet similar to the one at Fermilab with behavior comparable to what String theory is telling us. Let's assume that by inserting the muon in the magnet and holding it by a string, it passes through a circular clock with twelve energy indicators on it. The clock here has nothing to do with time; it is just an analogy for a circular tool with twelve indicators.

The twelve energy indicators correspond to the spectrum of energy qualities proposed earlier in this book and the key of 12 energy qualities covering this full energy spectrum, which we referred to as HDBR12. These are the spectrum of energies of the three imaginary complex forces as explained earlier in this book.

The idea here is that the proposed HDBR12 key implies that the muon will have either twelve or twenty-four different measurements of its g-factor, causing clockwise and anticlockwise rotation, which corresponds to the twelve energy qualities of the horizontal component of the wave on the proposed spectrum of energies causing clockwise rotation and potentially their opposite or complex *conjugates -HDBR12* of the vertical component of the wave causing anti-clockwise rotation*. This is, of course, not easy to detect and requires an extremely high level of accuracy of the experiment. Additionally, we will have to repeat the experiment many times to verify that we just get either twelve or twenty-four values for the g-factor no matter how many times we repeat the experiment.

It is expected that we will get more accurate results when designing a magnetic trap that has a round potential at the bottom, where the muons pass through a circular disc clock-like instrument. On this circular disc clock-like instrument, we could mark the energy spectrum of the HDBR12 and calibrate it for the energy we want to measure. This would be for measuring the horizontal component of the wave. We could potentially require a different setup to detect the vertical component of the wave, the –HDBR12*. We could possibly require two vertical circular disc clock-like with indications of the –HDBR12* where the HDBR12 would be sort of complex conjugated in the opposite direction. We can then calibrate it for the energy we want to measure.

There are chances that the muon at the moment of measuring is not interacting or carrying any resonance or operation with the imaginary complex forces. The thinking goes that in that case, there should neither be clockwise nor anticlockwise rotation for the muon. While if the muon gets in resonance with the HDBR12, we will get a clockwise rotation. If the muon gets in resonance with the –HDBR12*, we will get anticlockwise rotation.

So, the thinking goes that the muon will have twelve values for the clockwise rotation and twelve values for the anti-clockwise rotation. And various random values for a muon when it is not interacting with imaginary complex forces.

It is like antenna polarization, you do horizontal polarization to get the horizontal component of the wave and vertical polarization to get the vertical component of the wave. Other angles could still get you sound, but it will not be clear. So, we need to measure the clear values for the muon.

Suppose this idea turned out to be correct, or at least partially correct. Then this implies that the muon is in resonance with one of these twelve energy qualities and their "anti-qualities." It is not necessarily just resonance. It could be any of the operations of the proposed imaginary complex forces—either a single or complex operation.

The basic mathematics for the imaginary complex forces proposed in this book could be a start to track these twelve energies and their respective three forces. Another line of thought or more sophisticated and developed mathematics would help as well.

Could these operations and the idea of resonance with one of the energies of the HDBR12 be applied to all particles, waves, and fields, not only the muon?

The answer is probably yes; it could be applied to all particles, waves, and fields. The thing is that the muon is 200 times more massive than the electron. The probability of interaction between a particle and some massive virtual particle is proportional to mass squared. So, the muon is 40,000 times more likely than the electron to encounter a virtual Higgs boson, for example, or a virtual proton or other hadrons. It is 40,000 times more likely to encounter any unknown virtual particles. Yet, it is also expected that electrons and other particles and even their anti-particles —if we consider super symmetry- are regularly involved in resonance and other imaginary complex mother forces. However, if that is correct or even partially accurate, given the mass of the muon and the experimental tools we have, it is likely as well that we experimentally detect these types of resonance and other imaginary complex forces operations on the muons first.

So, we considered that the muon is held by a string. What about the length of the string? It could be viewed as a sort of secondary calibration that allows for more accuracy in detection. That would be, of course, if we figured out a way to adjust and calibrate the length of this imaginary string.

The graviton and the imaginary complex forces

The approach for testing string theory is mainly through particle collisions. Particle collisions allow all sorts of sub-atomic and "extraordinary" particles to be released. The string theory proposes that these collisions could generate the graviton giving evidence of gravity seeping of our membrane universe into "higher dimensions." These collisions should show that the amount of energy after the collisions are less than the amount of energy before the collisions as gravitons have seeped into higher dimensions. But because gravitons are so weak, a detector with almost the mass of Jupiter and 100% efficiency, placed in close orbit around a neutron star, would only be expected to observe one graviton every ten years, even under the most favorable conditions, which is pretty challenging.

If we want to examine if the graviton is escaping to higher dimensions by measuring the amount of energy before and after the collision, we might as well try to examine the effects of the imaginary complex forces introduced in this book on the graviton. The basic mathematics for the imaginary complex forces proposed in this book could be a start. Another line of thought or more sophisticated and developed mathematics would help as well. So, for example, we can try to measure the amount of resonance of the B force with the graviton after the collision. We can try to measure the amount of HD-entanglement of the HD force with the graviton. We can try to observe as well the creation and annihilation operations of the R-force on the graviton. The scaling and projection operations of the higher dimension force could be another candidate as well. Moreover, we can check the resonance with the HDBR12 energies with the graviton. These are just some proposals that would confirm the observation of the graviton and the imaginary complex forces if proven fully or at least partially correct.

We might need to design magnetic traps based on the approach explained above, with round potential at the bottom, a circular disk clock-like instrument, and different horizontal and vertical configurations.

Testability of the paradigm-shifting operation of the HD-force

The paradigm-shifting operation of the HD force refers to the operation that happens when the HD force operates on something or someone - either it operates on matter or organisms- it causes the acceleration of phase transition or even phase conversion. It changes the function of what it is operating on. This is not acceleration as the rate of change of velocity with respect to time; this is totally different, it is basically accelerating the phase transition or conversion by introducing new information to the operatee (The thing the HD force is operating on) or in other words to the wave function of the operatee. This information will lead to a paradigm shift in the operatee and a change in its function.

Some examples for the paradigm-shifting operations of the HD-force were explained in the Higher Dimension force section in this book. There could be potentially many ways to observe or carry out experiments to observe the paradigm-shifting operation of the HD force. Yet, of course, these experiments have to be reproducible.

One of the experiments worth mentioning was successfully "Using salt water for irrigation of fresh water plants." An experiment that implies sort of a paradigm shift that happened to a material "salt water" that led to a notable change in one of its functions could be a stray thread to be tugged for the idea of the paradigm shift operation of the Higher Dimension force.

This two months experiment, "Using salt water for irrigation of fresh water plants," was carried out in May 1998 by an Egyptian engineer named Adel Ammar to test the possibility of using Biogeometry to enhance fresh water plants to grow in salt water. Biogeometry is a concept introduced by Dr. Ibrahim Karim –an architect who is a graduate of the Federal Institute of Technology, Zurich, Switzerland, and teaches at several universities- using the energy principles of shape to qualitatively balance biological energy systems and harmonize their interactions with the environment. (At least this is how Ibrahim Karim explains it).

It is worth mentioning that the concepts of Biogeometry are neither widely recognized nor widely accepted. However, here we just want to check if this specific experiment, "Using salt water for irrigation of fresh water plants," could be of any help to test the idea of the paradigm-shifting operation of the proposed higher dimension force.

It is worth mentioning as well that even though the results of the experiment were positive and successful. The experiment was not widely recognized or accepted, probably because mainly it was not highly reproducible, and the mathematics behind it has not been clear in the first place.

Adel Ammar used water from the red sea of Sokhna in Egypt, which has a very high salinity, to plant sweet potatoes. A control potato was given fresh water, and another control potato was given salt water. For the third control potato, "Biogeometrical shapes" were used for the water container, the water channel, and the plant pot. The experiment lasted for two months.

The control potato that was given salt water shriveled by the end of the day, while the "Biogeometrically treated" one fared very well and budded normally. It fared slightly better with time than the one that was given fresh water, which started to show signs of decay.

As per the Biogeometry website, similar experiments with similar results were tried with other herbs planted in "Biogeometrically shaped" boundary and water containers. The Biogeometry website and book "Back to a future of mankind" also mention that several projects were done by Professor Peter Mols at the Agriculture University of Wageningen in Holand from 1999 until 2002, where "Biogeometry" was used to produce chemical-free apples. Again as per the Biogeometry`s website, the experiments showed a noticeable decrease of parasite infection due to the raised "immunity" of the plant. The apples had a longer shelf life. The yield per acre increased above average. However, unfortunately, these results were not published in any scientific papers that we know of, so it was not widely recognized or acknowledged.

In general, we can conclude from this experiment that certain configurations of shapes led to changes in energy. In other words, these could be changes in the information of the wave function, an update for the wave function of the system, or the field which leads to a paradigm shift. This could be the paradigm-shifting operation of the HD force proposed in this book.

Yet, two main things are required here to see if we can develop something from this line of thought—first, some equations/mathematics to track how this could have happened. Second, reproduced results. This simply means we will need to repeat the experiments many times and review if the results meet our expectations and predictions.

As explained earlier, the paradigm-shifting operation of the HD-force would be accelerating the phase transition or conversion by introducing new information to the operatee (The thing the HD force is operating on) or, in other words, to the wave function of the operatee. This information will lead to a paradigm shift in the operatee and a change in its function.

The proposed notation for this operation was:

$$\text{HD} \divideontimes f\hat{\Psi}(x,t) = \frac{\partial E}{\partial t}$$

- Where HD would be the higher dimension force.
- $\divideontimes$ is the "In sign" indicating that the HD force is operating "In" something or someone.
- $f\hat{\Psi}(x,t)$ Refers to a complicated field of wave functions in a specific space and time. With a note here, that space and time $f\hat{\Psi}(x,t)$ Are different from the space and time of the HD force. In other words, the HD force is totally independent of space and time parameters of the complicated field of wave functions $f\hat{\Psi}(x,t)$ it is operating "in."
- $\frac{\partial E}{\partial t}$ is just a notation to indicate the phase transition or conversation of the operatee (The thing the HD force is operating on).

This equation and notation are just indicative; it might contain mistakes, it might require some complex numbers, and Planck`s constant. It is just merely a basic description of what this paradigm-shifting operation might look like. A different notation or even a totally different mathematical framework could be used to describe the idea.

There is also a probability that the operation carried out on the salt water or the fresh water plants is not just the paradigm-shifting operation of the HD force; there could be other imaginary complex forces operations in play as well. The basic mathematics introduced in this book could be a start for a mathematical machine to track this operation; otherwise, a different mathematical framework could be used.

The paradigm-shifting operation of the HD force could be a very powerful toll if proven correct or even partially correct. This could assist us in changing the function of matter on some occasions if required, which will open a world of possibilities and applications.

Chapter 08:
The main new concepts introduced in this book

The main new concepts introduced in this book

1. **The idea of the Big Defreeze.**
   This thought experiment suggests that in the scenario of the Big Defreeze at the beginning of the universe, all matter started with a single collective quantum wave which could be referred to as the Identity Stem Wave of the Universe (ISWU), which is a form of a Bose-Einstein condensate. In other words, the universe was in a "frozen state" at almost absolute zero temperature and potentially shrunk in a very tiny collective wave. This is pretty much what is expected with a minor but fundamental difference which is that the condensed, tiny beginning of the universe was not hot; it was cold as it had no energy/heat, and then it got heated by the symmetry-breaking mechanism explained in this book.

   The Big Defreeze suggests as well that the universe started the same way it will end. The Big Defreeze is then compatible with the Big Rip and a minor modification on the Big Freeze. The book then suggests the source of the ISWU through the straight-line analogy.

2. **The identity stem waves (ISWs).**
   The book suggests that our universe started with sort of a reverse mechanism to the big freeze with a minor modification on that scenario. It started as a tiny collective quantum wave, which could be referred to as the **"Identity Stem Wave of the Universe" ISWU**. This stem wave started gaining energy leading to fluctuations behaving in the manner of a string forming what could be referred to as the "Identity stem waves of matter," leading to a super symmetrical crystal with no well-defined boundaries, where the "Identity stem waves" of all elements started forming and interacting. Until the identity stem wave of helium ISW_{He} broke the symmetry of the ISW crystal with speed comparable to the speed of light.

3. **Using the Fourier transform to understand the symmetry-breaking mechanism of the ISW crystal.**

   The frequencies of these identity stem waves ISWs are comparable to the temperature at which they form BEC. This thought experiment suggests using the Fourier transformation to test and measure the mechanism of forming the ISWs as spikes/fluctuations.

   It suggests that for the ISWs of all matter, the winding frequency is comparable to the ISW`s frequency except for helium; the winding frequency is expected to be 299,789,012 times the frequency of Helium`s ISW, which suggests a symmetry-breaking mechanism. This is a mechanism that breaks the symmetry of the super symmetrical crystal leading to an explosion or a big bang with a value comparable to the speed of light.

   In simpler words, this thought experiment suggests that our universe was frozen, and it got heated to a degree leading to the big bang, and it is heading towards getting frozen again.

4. **The straight-line analogy.**

   The analogy goes, what if the multiverse, the landscape of string theory, and all the information within it have a one-dimensional representation of reality as well. Suppose you consider the two-dimensional data and the three-dimensional data as two versions of reality or two reconstructions of the same reality. Then why do we not have a third version or the third reconstruction of reality in one dimension represented in a straight line?

   This line of thought would lead to the proposed source of the collective matter-wave of the universe which is referred to in this book as the identity stem wave of the universe ISWU. It would lead as well to the idea of the imaginary complex forces, their properties, and operations.

5. **Dealing with identity stem waves ISWs as strings.**
   Applying the mathematics of string theory on the idea of the identity stem waves ISWs.

6. **The imaginary complex forces and their potential properties.**
   Introducing three imaginary complex forces and their potential properties as a result of the idea of the Big Defreeze and the straight-line analogy. The three imaginary complex forces are:

   a. The Higher dimension force (The HD-force).
   b. The Balancing force (The B-force).
   c. The retrieving force (The R-force).

7. **Introducing the potential operations of the imaginary complex forces and the basics of their potential mathematics.**
   a. <u>The Higher Dimension force (The HD-force):</u> Scaling, projection, entanglement, and paradigm-shifting.
   b. <u>The Balancing force (The B-force):</u> Resonance.
   c. <u>The retrieving force (The R-force):</u> Creation and annihilation.

8. **The two-worlds interpretation.**
   The idea is that we have two worlds. One in our universe on that two-dimensional holographic film having quantum mechanical behavior, and the other one is totally outside "the notebook" (our multiverse) and its space-time parameters. It does not follow the quantum mechanical behavior. The version of you in our universe is an observer with constituents following the quantum mechanical behavior, which requires the second version of you in the other world with constituents that do not have quantum mechanical behavior.

9. **The mechanism of encoding information in the two-worlds interpretation.**
   A mechanism introduced in the two worlds interpretation, for the first version of you in our world/universe to be able to change the code which is reconstructed in different dimensions without any sort of interference or control either from the second version of you in the other world or from any other forces.

10. **The mechanism of updating the information of the wave function as per the two-worlds interpretation.**
    The two-worlds interpretation proposes a mechanism for updating the wave function of the "probability cloud," the other scenarios. Not by destroying the wave function or collapse of the wave function. It simply means updating the wave function in a sort of formatting all the other possibilities to either have an updated wave function with no information or with different information.

11. **Proposing a unification for the three complex forces, in addition to the potential properties and operation of this unification.**
    Proposing the unification of the imaginary three "Complex forces," the Higher dimension force (The HD-force), the Balancing force (The B-force), and the Retrieving force (The R-force), as one unified force, we will name it for short the $HDBR_3$.

    The analogy goes that we had the straight line as a one-dimensional representation of reality. A tiny fluctuation happened on that straight line –without changing its geometry- that was projected on another space forming the ISW crystal. When all matter is shattered to their elementary particles, form BECs, which start fluctuating, forming what is referred to as the identity stem waves until the winding frequency of Helium`s ISW causes symmetry breaking, leading to the big bang and inflation.

12. Applications of introducing the "In sign" to the Schrodinger Equation

$$i\hbar \frac{\partial}{\partial t}\psi(x,t) = \hat{E} \ast \hat{\psi}(x,t)$$

Here the equation is prescribing that the energy operator is acting "in" a wave function, which is the operatee in this case, and this wave function does not have definite energy; it could be a field of waves. This equation also prescribes that the energy operator and the wave operatee affect each other and can swap roles.

13. Identity stem waves ISWs, and the idea of energy in shape gives a specific function.

After the symmetry breaking, the ISWs produce a system of numbers, proportions, and angles. This system becomes the basis of forming shapes and matter. These shapes are an expression of all energy principles within them. This brings us to the idea that energy in shape gives a specific function.

$$\hat{E} \ast \hat{s} = f\hat{\Psi}(x,t)$$

Where $\hat{E}$ is the energy operator, $\hat{s}$ is the shape operatee and $f\hat{\Psi}(x,t)$ is the resulting wave function. This proposed equation prescribes that energy in shape gives a specific function/wave function, which could be applied in principle to almost everything.

The resulting wave function or "IN product" should be considered an operator. In this proposed equation, we have the freedom to swap the in "IN product," the operator, and the operatee. In other words, the energy operator could act on the wave function to give a specific shape or topological manifold, etc. in that case; the equation would be:

$$\hat{E} \ast f\hat{\Psi}(x,t) = \hat{s}$$

Swapping the roles of the operator, operatee, and the "IN product" is a fundamental principle that could be useful in many conditions and various states.

14. The potential collective operations of the unified imaginary complex forces.

Proposing some of the potential collective operations of the unified imaginary complex forces. In addition to the potential basic mathematics and notations for these operations.

$$\text{HDBR3 } \phi(x) \, \Gamma_{ij}^k \, a_n^+ \divideontimes \hat{s} \, f\widehat{\Psi}(x,t) = \sum_{HDBR3}^{n} \divideontimes |S_b\rangle \, f\widehat{\Psi}(x,t)(\sigma : SP_m \longrightarrow SP_n)$$

Where:
- HDBR3 is the unified complex forces
- $\phi(x)$ is the scaling and projection operator
- Γ_{ij}^k is the connection coefficient
- a_n^+ is the creation operator, it could be an annihilation operator a_n^-
- $\divideontimes$ is the sign indicating that it is operating "in" something.
- $\hat{s}$ is the shape of the operatee, which affects the operation (the balancing force could lead to a change in the shape and geometry of the operatee).
- $f\widehat{\Psi}(x,t)$ is a field of waves or wave function the B-force is operating "in"
- $\sum_{HDBR3}^{n} \divideontimes$ is the sum of the spectrum of energies of the HDBR$_3$ on the left side of the operation and acting "in" the resulting field of wave/wave function.
- $|S_b\rangle$ refers to the balanced state.
- $f\widehat{\Psi}(x,t)$ is the resulting field of waves/wave function
- σ is a point or section on the base space
- SP_m is the main space
- SP_n is the nth space

What this equation is telling us is that the unified HDBR$_3$ forces, the Higher dimension force (The HD-force), the Balancing force (The B-force), and the retrieving force (The R-force) are acting collectively on a field of waves/wave function using their mechanisms and operators including scaling and projection, connection, resonance and transformation, creation and annihilation. The shape of the operatee could affect and get affected by the operation. The resultant is the sum of the spectrum of energies resulting from the operation acting "in" an updated field; this field could be in a balanced state. The operation would lead to scaling and projection of a point or a section from the main space (it could be another space, not necessarily the main space) on the base space (or any other space). The equation is highly interchangeable, and some of the operators could be isolated depending on the operation. Planck's constant might be required to be introduced in some cases.

15. **The mechanism of the HD-force`s entanglement.**
    The nature of the entanglement properties of the higher dimension force (The HD-force) is not necessarily always as quantum entanglement. Potentially for the mechanism of HD-force`s entanglement:
    - The HD force`s entanglement will not be limited to a specific number of states.
    - "Partial entanglement" could be allowed, which means that the two entangled systems could be partially entangled, in a sense that some of the properties of one of the systems or fields could be entangled with some of the properties of the other system or field. And not necessarily the entire system/particle. Entanglement entropy could be the measure of the degree of entanglement.
    - "Fading of entanglement" could be allowed and tracked as well. The term fading of entanglement would probably be more accurate than using "decay of entanglement."
    - The HD-force could choose to pass the information to an imaginary particle/field, which is not limited by our space-time parameters, and then this particle/field would be entangled with the particle/field defined by our space-time parameters.

16. **The notebook thought experiment.**
    A thought experiment that helps in understanding the two-worlds interpretation and the operations of the proposed imaginary complex forces.

17. **The spectrum of energies of the $HDBR_3$**
    A key of 12 energy qualities covers the entire energy spectrum. Let's call this key HDBR12. We will consider that the wave has a horizontal component and a vertical component. The HDBR12 would represent the horizontal component of the wave. It would have opposite direction energies, representing the vertical component of the wave; we can use the notation the $-HDBR12^*$ to represent it.

18. **The paradigm-shifting operation of the HD-force**
    This means that when the HD force operates on something or someone -either it operates on matter or organisms- it causes the acceleration of phase transition or even phase conversion. It changes the function of what it is operating on. This is not acceleration as the rate of change of velocity with respect to time. This is totally different and could have different mathematics describing it as well; this is basically accelerating the phase transition or conversion by introducing new information to the operatee (The thing the HD force is operating on) or, in other words, to the wave function of the operatee. This information will lead to a paradigm shift in the operatee and a change in its core function.

19. The balancing force and the fine-tuning

The balancing force along with the other proposed imaginary complex forces, could give a plausible explanation for the fine-tuning from a different perspective, which would simply be that the universe is in a "balanced state" rather than in a fine-tuned state. This balanced state is driven by the balancing force for a specific period of time and space to achieve specific conditions for different forms of life generation mechanisms.

There would be another operating force here, which would be the retrieving force which would be working on retrieving this kind of ecosystem to its original state, and this is yet another important differentiation between the balancing force and the retrieving force.

Bibliography and references

Chapter 01: The Big Defreeze

1. Physics 12c, Statistical Mechanics, Spring 2016, Caltech, Course description: **An introductory course in statistical mechanics.** Instructor: John Preskill, http://theory.caltech.edu/~preskill/ph12c/
2. Leonard Susskind, Stanford, **String theory & M-theory**, Lecture 02, September 2010.
3. **Three ways the universe could end** - Venus Keus, TED-Ed Youtube Channel, Feb 19, 2019.
4. Katie Mack, **The End of Everything: (Astrophysical speaking)**, Scribner Book Company, Press, 04 May 2021.
5. 2001 Nobel Laureate Lecture in Physics - Wolfgang Ketterle, **The Story of Bose-Einstein Condensates**, MIT Video Productions YouTube channel, Apr 14, 2018.
6. **The First Three Minutes: A Modern View Of The Origin Of The Universe**, Steven Weinberg, Basic Books; 2nd Updated ed. edition (18 August 1993).
7. **Fermions to bosons, bosons to fermions**, ScienceDaily, March 14, 2016. https://www.sciencedaily.com/releases/2016/03/160314111135.htm
8. Okniński, A. **On the Mechanism of Fermion-Boson Transformation**. Int J Theor Phys 53, 2662–2667 (2014). https://doi.org/10.1007/s10773-014-2062-4
9. **A Fermi gas of atoms**, physicsworld, April 4, 2002.
10. **The Fermion-Boson transformation in fractional quantum Hall systems,** John J. Quinn, Arkadiusz Wojis, Jennifer J. Quinn, Arthur T. Benjamin, Physica E 9 (2001) 701-708 http://math.hmc.edu/wp-content/uploads/sites/5/2019/06/Fermion-Boson-Transformation-in-Fractional-Quantum-Hall-Systems.pdf
11. **But what is the Fourier Transform? A visual introduction**, 3Blue1Brown YouTube channel, Jan 26, 2018.
12. **WMAP – Fate of the Universe,** WMAP's Universe, NASA. July 17, 2008.
13. Krauss, Lawrence M.; Starkman, Glenn D. (2000). **"Life, the Universe, and Nothing: Life and Death in an Ever-expanding Universe."** Astrophysical Journal. 531 (1): 22–30. arXiv:astro-ph/9902189.

14. Kirshner, Robert P. (13 April 1999). **"Supernovae, an accelerating universe and the cosmological constant."** Proceedings of the National Academy of Sciences.
15. **Inflation**, Alan H. Guth, Massachusetts Institute of Technology, https://sites.astro.caltech.edu/~ccs/Ay21/guth_inflation.pdf

Chapter 02: The straight-line analogy

16. Alan Guth, *Inflationary Cosmology: Is Our Universe Part of a Multiverse? Part I*, MIT OpenCourseWare, MIT 8.286 The Early Universe, Fall 2013.
17. Alan Guth. 8.286 **The Early Universe**. Fall 2013. Massachusetts Institute of Technology: MIT OpenCourseWare, https://ocw.mit.edu. License: Creative Commons BY-NC-SA. https://ocw.mit.edu/courses/physics/8-286-the-early-universe-fall-2013/#
18. Frederick Denif, *String theory Vacua*, "String Theory and its Applications: from meV to the Plank Scale" held at the Theoretical Advanced Study Institute, Jun01-25, 2010.
19. **String Theory Landscape** - Alexander Westphal (SETI Talks), SETI Institute YouTube Channel, May 04, 2012.
20. **Topological manifolds and manifold bundles**- Lec 06 - Frederic Schuller, from a series of lectures - "Lectures on the Geometric Anatomy of Theoretical Physics, Fredric Schuller YouTube Channel, Sep 22, 2015.
21. Leonard Susskind on **The World As Hologram,** TVO Docs YouTube Channel, Nov 4, 2011.
22. **Can a Circle Be a Straight Line?** | Space Time | PBS Digital Studios, PBS Space Time YouTube channel, Jul 2, 2015.
23. **General Relativity & Curved Spacetime Explained!** | Space Time | PBS Digital Studios, PBS Space Time YouTube channel, Jul 30, 2015.
24. John Preskill, Caltech, 2011, ph12c lecture 11 **BEC (Bose-Einstein Condensation)**
25. Frederik Denef - **Vacua** - Lecture 3, Lecture at the 2010 TASI summer school on "String Theory and its Applications: from meV to the Plank Scale" held at the Theoretical Advanced Study Institute, Jun01-25, 2010. GraduatePhysics YouTube channel, Jun 4, 2016.
26. Lecture 2: **Topological Manifolds** (International Winter School on Gravity and Light 2015), Frederic Schuller, The WE-Heraeus International Winter School on Gravity and Light YouTube channel, Feb 17, 2015.

27. Particle Physics (2018) Topic 16: **The Higgs Mechanism and Spontaneous Symmetry Breaking,** Alex Flournoy, Colorado School of Mines, Alex Flournoy Youtube channel, Mar 9, 2018.
28. Particle Physics Lecture 15: **The Higgs Mechanism,** Alex Flournoy, Colorado School of Mines, Alex Flournoy YouTube channel, Feb 28, 2020.
29. Particle Physics Lecture 16: **The Higgs Mechanism and Spontaneous Symmetry Breaking,** Alex Flournoy, Colorado School of Mines, Alex Flournoy YouTube channel, Mar 4, 2020.

Chapter 03: ISWs as strings

30. Leonard Susskind, Stanford, **Lecture 4 | String Theory and M-Theory,** October 11, 2010, Stanford YouTube Channel.
31. Leonard Susskind, Stanford, **Lecture 2 | String Theory and M-Theory,** September 27, 2010, Stanford YouTube Channel.
32. Leonard Susskind, Stanford, **Lecture 2 | New Revolutions in Particle Physics: Standard Model,** January 18, 2010, Stanford YouTube Channel.
33. Leonard Susskind, Stanford, **Lecture 3 | String Theory and M-Theory,** October 04, 2010, Stanford YouTube Channel.
34. Professor Dave Explains, **Quantum Numbers, Atomic Orbitals, and Electron Configurations,** Professor Dave Explains YouTube Channel, August; 2015.
35. Tensor Calculus 22: **Riemann Curvature Tensor Geometric Meaning (Holonomy + Geodesic Deviation),** eigenchris YouTube channel, Jun 17, 2019.
36. Tensor Calculus 24: **Ricci Tensor Geometric Meaning (Sectional Curvature),** eigenchris YouTube channel, Oct 14, 2019.
37. Tensor Calculus 25 - **Geometric Meaning Ricci Tensor/Scalar (Volume Form),** eigenchris YouTube channel, Oct 25, 2019.
38. **Fundamental Lessons From String Theory** – ICTP Colloquium, By Prof. Cumrun Vafa, Int'l Centre for Theoretical Physics YouTube channel, Mar 1, 2017.

Chapter 04: The imaginary complex forces

39. **Electroweak Theory and the Origin of the Fundamental Forces,** PBS Space Time YouTube channel, Nov 04, 2020.
40. **The Four Fundamental Forces of nature - Origin & Function,** Arvin Ash YouTube channel, Jul 10, 2020.

41. **Why & How do the 4 fundamental forces of nature work?** Arvin Ash YouTube channel, Jul 18, 2020.
42. **First results from Fermilab's Muon g-2 experiment strengthen evidence of new physics**, Fermilab, April 7, 2021. https://news.fnal.gov/2021/04/first-results-from-fermilabs-muon-g-2-experiment-strengthen-evidence-of-new-physics/
43. **Have Scientists Really Discovered a New FORCE? Muon g-2 Experiment** EXPLAINED by Parth G, Parth G YouTube channel, April 13, 2021.
44. **Muon g-2 experiment finds strong evidence for new physics**, Fermilab YouTube channel, April 7, 2021.
45. Barrow, John D. (2002), **The Constants of Nature**; From Alpha to Omega – The Numbers that Encode the Deepest Secrets of the Universe, Pantheon Books, ISBN 978-0-375-42221-8.
46. P. R. Bunker; I. M. Mills; Per Jensen (2019). **"The Planck constant and its units"**. J Quant Spectrosc Radiat Transfer. 237: 106594.
47. Chang, Donald C. (2017). **"Physical interpretation of Planck's constant based on the Maxwell theory"**. Chin. Phys. B. 26 (4): 040301. arXiv:1706.04475. doi:10.1088/1674-1056/26/4/040301.
48. **Planck's Constant and The Origin of Quantum Mechanics** | Space Time | PBS Digital Studios, PBS Space Time YouTube channel, Jun 22, 2016.
49. Ian Morison (2008). **Introduction to Astronomy and Cosmology**. J Wiley & Sons.
50. Tomokazu Kogure; Kam-Ching Leung (2007). "2.3: **Thermodynamic equilibrium and black-body radiation**". The astrophysics of emission-line stars. Springer.
51. Cole, George H. A.; Woolfson, Michael M. (2002). **Planetary Science: The Science of Planets Around Stars** (1st ed.). Institute of Physics Publishing.
52. Michio Kaku: **What's the Fate of the Universe? It's in the Dark Matter** | Big Think, Big Think YouTube Channel, Nov 08, 2012.
53. **The Theory of (almost) Everything Explained (almost) Intuitively**, Arvin Ash YouTube channel, May 01, 2021.
54. **Dark Matter: The Situation has Changed**, Sabine Hossenfelder YouTube channel, May 01, 2021.
55. Dimitri Nanopoulos, **What is a particle?,** Quanta Magazine, Editor: Natalie Wolchover, November; 2020.
56. Mohammad Hasan Algarhy, **Quantizing Emotions (The Basic Mathematics of Psychology)**, Feb 17, 2021.

57. Barton Zwiebach, *Operators and the Schrödinger Equation*, MIT 8.04 Quantum Physics I, spring 2013, MIT OpenCourseWare, License: Creative Commons BY-NC-SA.
58. Wolfgang Ketterle, 1. **Resonance I**, MIT OpenCourseWare YouTube channel, March 24, 2015.
59. The Nobel Prize in Physics 2001, Wolfganag Ketterle, https://www.nobelprize.org/prizes/physics/2001/ketterle/facts/
60. Wolfgang Ketterle. 8.421 **Atomic and Optical Physics** I. Spring 2014. Massachusetts Institute of Technology: MIT OpenCourseWare, https://ocw.mit.edu. License: Creative Commons BY-NC-SA. https://ocw.mit.edu/courses/physics/8-421-atomic-and-optical-physics-i-spring-2014/#
61. Allan Adams, Matthew Evans, and Barton Zwiebach. 8.04 **Quantum Physics** I. Spring 2013. Massachusetts Institute of Technology: MIT OpenCourseWare, https://ocw.mit.edu. License: Creative Commons BY-NC-SA.
62. **Is the Universe Fine Tuned for Life? The Case FOR and AGAINST Fine Tuning, Arvin Ash YouTube channel**, May 15, 2021.
63. **Leonard Susskind - Is the Universe Fine-Tuned for Life and Mind?** Closer To Truth YouTube channel, Jan 8, 2013.
64. "Fine-Tuning." The Stanford Encyclopedia of Philosophy. Center for the Study of Language and Information (CSLI), Stanford University. Aug 22, 2017.
65. Gribbin. J and Rees. M, **Cosmic Coincidences: Dark Matter, Mankind, and Anthropic Cosmology** p. 7, 269, 1989, ISBN 0-553-34740-3.
66. University of New South Wales. "**New findings suggest laws of nature not as constant as previously thought.**" ScienceDaily. ScienceDaily, 27 April 2020. <www.sciencedaily.com/releases/2020/04/200427102544.htm>.
67. Michael R. Wilczynska, John K. Webb, Matthew Bainbridge, John D. Barrow, Sarah E. I. Bosman, Robert F. Carswell, Mariusz P. Dąbrowski, Vincent Dumont, Chung-Chi Lee, Ana Catarina Leite, Katarzyna Leszczyńska, Jochen Liske, Konrad Marosek, Carlos J. A. P. Martins, Dinko Milaković, Paolo Molaro, Luca Pasquini. **Four direct measurements of the fine-structure constant 13 billion years ago**. Science Advances, 2020; 6 (17): eaay9672 DOI: 10.1126/sciadv.aay9672.
68. **Are the laws of nature changing with time?** physicsworld, April 1, 2013. https://physicsworld.com/a/are-the-laws-of-nature-changing-with-time/

69. Duff, M. J. (2014). **"How fundamental are fundamental constants?** "Contemporary Physics. 56 (1): 35–47. arXiv:1412.2040. doi:10.1080/00107514.2014.980093 (inactive 2021-01-17).
70. Georgescu, I. **A changing constant?**. Nature Phys 13, 824 (2017). https://doi.org/10.1038/nphys4261
71. Paschotta, Rüdiger (2008). **Encyclopedia of Laser Physics and Technology**, Vol. 1: A-M. Wiley-VCH. p. 580. ISBN 978-3527408283.
72. **Obliquity (change in axial tilt),** Earth Observatory, Nasa, Mar 24, 2000. https://earthobservatory.nasa.gov/features/Milankovitch/milankovitch_2.php#:~:text=Today%2C%20the%20Earth's%20axis%20is,between%2022.1%20and%2024.5%20degrees.
73. Lectures 1 and 2 of Leonard Susskind's course concentrating on **Quantum Entanglements** (Part 1, Fall 2006). Recorded September 25, 2006 at Stanford University.
74. **The Vacuum Catastrophe**, PBS Space Time YouTube channel, Nov 03, 2017.
75. Ibrahim Karim, **Back to a future for mankind**, Biogeometry consulting Ltd, March, 2010.
76. Adler, Ronald J.; Casey, Brendan; Jacob, Ovid C. (1995). "**Vacuum catastrophe: An elementary exposition of the cosmological constant problem**". American Journal of Physics. 63 (7): 620–626. Bibcode:1995AmJPh..63..620A. doi:10.1119/1.17850. ISSN 0002-9505.
77. "A simple explanation of mysterious space-stretching 'dark energy?'". Science | AAAS. 10 January 2017.
78. Pagels, Heinz R. (2012). **The Cosmic Code: Quantum Physics as the Language of Nature**. Courier Corp. pp. 274–278. ISBN 9780486287324.

Chapter 05: The two worlds interpretation

79. **The Many Worlds of the Quantum Multiverse** | Space Time | PBS Digital Studios, PBS Space Time YouTube channel, Oct 27, 2016.
80. Everett, Hugh; Wheeler, J. A.; DeWitt, B. S.; Cooper, L. N.; Van Vechten, D.; Graham, N. (1973). DeWitt, Bryce; Graham, R. Neill (eds.). **The Many-Worlds Interpretation of Quantum Mechanics**. Princeton Series in Physics. Princeton, NJ: Princeton University Press. p. v. ISBN 0-691-08131-X.
81. Vaidman, L. **"Probability in the Many-Worlds Interpretation of Quantum Mechanics."** In: Ben-Menahem, Y., & Hemmo, M. (eds), The Probable and the Improbable: Understanding Probability in Physics, Essays in Memory of Itamar Pitowsky. Springer.

82. **The Trouble with Many Worlds**, Sabine Hossenfelder YouTube channel, Sep 27, 2019.
83. **Why the multiverse is religion, not science**, Sabine Hossenfelder YouTube channel, Jul 9, 2019.
84. **Topological manifolds and manifold bundles**- Lec 06 - Frederic Schuller, from a series of lectures - "Lectures on the Geometric Anatomy of Theoretical Physics, Fredric Schuller YouTube Channel, Sep 22, 2015
85. **The Problem with Quantum Measurement**, Sabine Hossenfelder YouTube channel, Oct 22, 2019.
86. **Understanding Quantum Mechanics #5: Decoherence**, Sabine Hossenfelder YouTube channel, Aug 15, 2020.
87. Lecture 5 of Leonard Susskind's course concentrating on **Quantum Entanglements** (Part 1, Fall 2006). Recorded October 23, 2006, at Stanford University.

Chapter 06: The unified imaginary complex forces

Chapter 07: Testability of the ideas introduced in this book

88. 2001 Nobel Laureate Lecture in Physics - Wolfgang Ketterle, **The Story of Bose-Einstein Condensates**, MIT Video Productions YouTube channel, Apr 14, 2018.
89. **Do Complex Numbers Exist?**, Sabine Hossenfelder YouTube channel, Mar 6, 2021.
90. **Quantum physics needs complex numbers**, Marc-Olivier Renou, David Trillo, Mirjam Weilenmann, Le Phuc Thinh, Armin Tavakoli et al. (Jan 26, 2021) e-Print: 2101.10873 [quant-ph] https://arxiv.org/abs/2101.10873
91. **Agriculture Research Projects, Biogeometry**, Ibrahim Karim, https://www.biogeometry.ca/biogeometry-agriculture-research-project
92. John Preskill, Caltech, 2011, ph12c lecture 11 **BEC (Bose-Einstein Condensation).**
93. LondonCityGirl, *String Theory Made Simple*, LondonCityGirl YouTube Channel, January; 2018.
94. **First results from Fermilab's Muon g-2 experiment strengthen evidence of new physics**, Fermilab, April 7, 2021. https://news.fnal.gov/2021/04/first-results-from-fermilabs-muon-g-2-experiment-strengthen-evidence-of-new-physics/
95. **String Theory Landscape** - Alexander Westphal (SETI Talks), SETI Institute YouTube Channel, May 04, 2012.

96. Lecture 2 of Leonard Susskind's course concentrating on **Quantum Entanglements** (Part 1, Fall 2006). Recorded October 2, 2006, at Stanford University. Stanford YouTube channel.
97. **Quantum Theory of Condensed Matter**, John Chalker, Physics Department, Oxford University, 2013 https://www-thphys.physics.ox.ac.uk/people/JohnChalker/qtcm/lecture-notes.pdf

Index

A

Absolute zero, 17, 18, 35, 161, 164, 184.

Analogy, 12, 23, 39, 44-49, 57, 60,-63, 66, 74, 78, 83, 84, 91, 98-99, 101, 104, 108, 114-118, 121, 132, 149, 164, 170, 181, 183.

Angular momentum, 36, 57, 72.

Anomalous magnetic moment, 77.

Asymmetry of matter, 105.

B

Balanced state, 99-103, 106, 153-158, 186, 188.

Balancing Force, 78, 84, 91-106, 118, 135, 149-158, 185,188.

Big Bang, 11, 12 18-24, 39, 40, 48, 63, 120, 127, 148, 150, 161, 164, 165, 182, 184.

Big Bounce, 20, 23.

Big Crunch, 11, 19-22, 78.

Big Defreeze, 17-37, 78, 119, 149, 160-167, 180, 182.

Big Freeze, 11, 17-22, 46, 63, 78, 160, 164, 180.

Big Rip, 11, 17-22, 35, 78, 260, 180.

Born rule, 130.

Bose-Einstein Condensates, 12, 17-21, 24-35, 49, 160, 166, 180.

Bosons, 12, 17, 34, 35, 74, 161, 166, 167.

C

Cave man analogy, 121-123.

Coherence, 126, 165, 166.

Cold atom lab, 18, 162.

Complex numbers, 14, 108, 144, 168, 178.

Constants of nature, 78, 91, 103-106.

Cosmic Microwave Background,

Cosmological constant, 30.

D

Dark energy, 11, 19, 21, 30, 75, 78, 80, 81, 105, 163.

Dark matter, 30, 105, 163.

Decoherence, 125, 130, 131, 141, 142.

Dirichlet boundary conditions, 63.

E

Electromagnetic force, 20, 40, 77, 147, 148.

Electromagnetic waves, 25, 32.

Electroweak energy, 101.

Electrons, 32, 70, 81, 96, 98, 124, 173.

Energy in shape, 69, 79, 85, 88, 89, 153, 185.

Entanglement, 72, 81, 115-118, 125, 134-144, 150-155, 168-170, 174, 186.

Entanglement-swapping, 170, 171.

Expansion of the universe, 11, 19-21, 30, 78, 101, 104, 163, 164.

F

Fermions, 12, 17, 34-37, 161, 164, 166.

Fermion-Boson transformation, 36, 162, 165.

Fermions problem, 12, 17, 34-37.

Fine tuning, 101, 103, 106, 188.

Fourier transform, 47, 48, 93, 163, 182.

Fundamental forces, 13, 21, 24, 34, 65, 77, 83, 147, 153.

G

Graviton, 66, 174.

Gravity, 11, 14, 19, 20, 28, 40, 43, 75, 77-80, 112, 147, 174.

H

Hadrons, 77.

$HDBR_3$, 68, 109, 117, 136, 148-159, 184, 186, 188.

HDBR12, 151, 157, 172-175, 188.

-HDBR12*, 151, 173, 188.

Higher dimension force, 13, 45, 78, 84, 100, 107-118, 135, 137, 149, 151, 154, 155, 157, 174-178, 182, 183.

Higgs Boson, 78, 149, 174.

I

Identity stem waves ISWs, 12, 17-19, 22-27, 29-35, 47-75, 102, 111-114, 150, 156, 161-168, 181-188.

Identity stem wave of the universe ISWU, 12, 17, 22, 23, 29, 35, 51, 52, 161, 181, 182.

Identity stem wave of helium, ISW_{He}, 25, 27, 56, 163, 164, 181.

Identity stem wave of iron ISW_{Fe}, 27, 30, 31, 51, 67, 111, 163, 164, 166.

Imaginary complex forces, 13, 14, 44, 68, 76-120, 135-145, 154, 164, 168-174, 178, 181-188.

Inflation, 39, 40, 104, 120, 149, 150, 161, 163-165, 184.

In sign, 68, 85-89, 107, 110, 136, 152-154, 178, 184.

Interference of BECs, 28, 32, 168.

L

Large Hadron collider, 18.

Laser cooling, 27-32, 164.

Landscape of string theory, 39, 41, 44, 45, 112, 182.

Leonard Susskind, 104, 105.

M

Magnetic trap, 27, 172, 175.

Many worlds interpretation, 13, 125, 127, 133, 134, 142.

Matter waves, 24-35,

Measurement problem, 131-139, 143.

Multiverse, 13, 44, 45, 104, 123, 133, 145, 182, 183.

Muon, 77, 81-84, 171-174.

N

Nano-kelvin temperatures, 25, 27, 83.

Neumann boundary conditions, 62, 63.

Notebook thought experiment, 117-120.

O

Optical trap, 33, 34, 166.

Optical lattice, 32, 34, 166.

P

Paradigm shifting operation, 108, 109, 176-179.

Patterns, 28, 70, 71, 82, 132, 143.

Q

Quantum entanglement, 72, 116, 117, 137, 139-145, 187.

Quantum fluctuations, 22, 163.

Quantum measurement, 131.

R

Resonance, 72, 92-95, 115, 137, 145, 156-159, 171-175.

Retrieving force, 60, 64, 75, 80-91, 118, 136, 150, 152-155, 184, 185, 188.

S

Sentient puddle, 105-107.

Scale-invariance, 166.

Scaling & projection, 109-111, 187.

Schrödinger's equation, 87, 88.

Schrödinger's cat, 126.

Spin, 72, 77, 81, 83, 97, 152, 157.

Straight line analogy, 74, 100-119, 136, 139, 150-152, 181-183.

Symmetry breaking, 12, 25, 35, 48, 56, 74, 109, 150, 165, 181, 182, 185.

T

Temperature fluctuations, 26, 30, 163.

Two worlds interpretation, 124-146, 183, 184.

U

Uncertainty principle, 58.

Unified imaginary complex forces, 147-159.

W

Wolfgang Ketterle, 24, 27, 28, 33, 92, 167.

The Big Bang theory predicts that all matter, time, and space began at an incredibly tiny compact state about 13.7 Billion years ago. This initial state was a hot, dense, uniform "soup" of particles that filled space uniformly and was expanding rapidly.

The Big Defreeze suggests a set of analogies and thought experiments about how the universe could have started and the state that preceded the Big Bang and cosmic inflation.

One of the analogies used in the book is called the straight line, which suggests a one-dimensional representation of reality comparable to the two-dimensional representation used in the holographic principle. The straight-line analogy proposes the reason behind the first fluctuation before the Big Bang. Additionally, it implies the existence of three forces referred to as the imaginary complex forces explained in detail within this book.

The notebook thought experiment and the two-worlds interpretation are other concepts introduced in this book that propose a solution to quantum mechanics' measurement problem along with a different perspective for the many-worlds interpretation.

The book suggests many analogies and thought experiments dealing with the beginning of the universe, fine-tuning, the end of the universe, the many world interpretations, entanglement, and others, including the proposed testing methods for these ideas.

The book proposes a mathematical framework and notations to describe the ideas mentioned in the book, which will require further scientific research and validation along with all the concepts and analogies proposed. A different mathematical framework could be used as well.

www.ingramcontent.com/pod-product-compliance
Lightning Source LLC
Chambersburg PA
CBHW060832220526
45466CB00003B/1076